On Socialization
in Hamadryas Baboons

On Socialization in Hamadryas Baboons

A FIELD STUDY

Jean-Jacques Abegglen

LEWISBURG
BUCKNELL UNIVERSITY PRESS
LONDON AND TORONTO: ASSOCIATED UNIVERSITY PRESSES

Associated University Presses
440 Forsgate Drive
Cranbury, NJ 08512

Associated University Presses
25 Sicilian Avenue
London WC1A 2QH, England

Associated University Presses
2133 Royal Windsor Drive
Unit 1
Mississauga, Ontario
Canada L5J 1K5

Library of Congress Cataloging in Publication Data

Abegglen, Jean-Jacques, 1945–
    On socialization in hamadryas baboons.

    Originally presented as the author's thesis (Ph.D.—
University of Zurich, 1976)
    Bibliography: p.
    Includes index.
    1. Hamadryas baboons—Behavior.   2. Social behavior
in animals.   I. Title.
QL737.P93A26   1984   599.8'2   80-70316
ISBN 0-8387-5017-6

Printed in the United States of America

To my parents

Contents

List of Figures

11

List of Tables

13

Preface

In the early years, the universe of vertebrate social ethology was typically restricted to the interaction between two animals. Research was concerned with the motivations and key stimuli that fired these interactions, with imprinting, and sometimes with dyadic bonds between mates or mother and young. Beyond the dyad, societies were described in terms of dominance order and the general behavioral tendencies of sex-age classes. In the past fifteen years, however, field workers and particularly primatologists have been privileged to learn of species in which organized social life extends far beyond the bond of two. They have discovered vertebrate social structure.

Naturalists had for a long time known and admired the complex societies of termites, bees, and ants. These societies are huge families. Their members care for each other by a division of labor so profound that it is imprinted in their morphology. Some members of the same colony look as different as if they belonged to several species. These societies are nevertheless anonymous. They have castes but no subgroups whose members recognize each other as individuals.

Complex vertebrate societies are the very opposite. Division of labor is modest; adult members look strikingly different only if they are male and female; and complex social structures can exist only because enough members have learned enough others by heart, as individuals. In the 1960s and 1970s, field students found that a troop of a hundred heads quite often is neither the chaotic anonymous horde that it seems, nor the single line of rank holders as the dominance cliché would have it. Animals such as the classical prairie dogs were found to form subgroups within groups, which all had somehow to be represented in the brains of at least some members, since strangers were recognized and treated differently. The term *hierarchy*, which had been misused for dominance orders, was found to apply far better to the stratified structure of these societies.

Among primates, no species displays a greater talent for grouping and subgrouping than the hamadryas baboon of eastern Ethiopia. During the first Ethiopian field study, described in my monograph *Social Organization of Hamadryas Baboons* (1968), we were immediately faced with three levels of hamadryas society: the great numbers of baboons that converged on the same vertical cliff among the semidesert hills of the south-

15

ern Danakil were certainly "a group," since they proceeded to settle down in the inaccessible ledges at dusk and passed the night as a "unit," with no other baboons for four miles around. However, the numbers found on the same cliff continued to vary from day to day; the "troops," as we call these cliff aggregations, were found to be no more than users of the same cliff on a particular night, who parted without further ado the next morning. Their size seemed to depend merely on how many other cliffs were available in the neighborhood.

True and stable social units were found only within the troop. Regular counts indicated constant "bands" as recurrent components of variable troops at several cliffs. Not all bands were on cliff-sharing terms; details of the hostile encounters between less familiar bands suggested that band males cooperate in the defense of their females against their opponents. Within the band, however, each adult female is exclusively bonded for years with her male; that is, the individual male leads and herds one or more females as his "one-male unit," and he defends his females against males of his own band as well. Leaders of one-male units are therefore cooperators as well as competitors. Experiments by Kummer, Götz, and Angst (1974) demonstrated an inhibition that reduces the competition among rival males within the band. We also showed experimentally that the males enhance the cohesion of their unit by aggressively herding their straying females.

Abegglen's *On Socialization* is the twin monograph to my *Social Organization*. First, he and his wife Helga recognized and demonstrated a fourth level of hamadryas organization: the "clan", located between one-male units and bands. He then goes beyond the description and short-term causal analysis and undertakes the study of how individuals, in the course of their life, "flow" through the social structure: how they move from one subgroup to another, recruit females and followers, and eventually lose them again in old age. He investigates hamadryas society in terms of individual life cycles, and he succeeds. The position of the subadult male follower to a one-male unit, known but not understood in the earlier studies, emerges as a waiting position for the takeover of a juvenile daughter. This process is found to differ consistently among bands: the daughters are robbed in one band and patiently courted in another. The Abegglens also witnessed a dramatic generation turnover in which three mature unit leaders lost their females to the younger generation of males. Females were not reassigned at random; rather, they appear as a kind of heritage passed mainly from older to younger males within an age-graded male unit, the clan.

This finding is the unique contribution of the study to our knowledge of primate societies: in other baboons and among macaques the stable frame of the reproductive unit is the membership of females. The rhesus groups of Cayo Santiago are composed of matrilines. Dominance and grooming networks are patterned by lifelong mother-daughter relationships, while males tend to move between groups. The history of Band I at Cone Rock as described by Abegglen shows the reverse: males remain in their clans for life, while females can be transferred. This implies that clan males are kin. In fact, hamadryas males seem to superimpose a set of patrilines onto the ancient matrilineal structure. Abegglen's results should be particularly interesting to anthropologists. Whereas chimpanzees are most close to humans in terms of tool use and cognition, hamadryas baboons may resemble us more in social structure: the society at Cone Rock has remarkable similarities with the segmental lineage system of the very nomad people who share their subdesert habitat.

Abegglen interprets the structures and processes of hamadryas by a systematic application of my bonding-stage concept of 1975 after confirming its basic phenomena in the developmental analysis of interactions. He searches and finds the subtle antecedents of important events in the histories of relationships and finally builds his synthesis on a rich and finely meshed web of data. While he cannot substantiate every strand of his final framework, his synthesis opens a new view on social structure as a landscape of dyadic relationships, founded years ahead of their productive phases, and affecting one another according to rules that only now begin to appear.

After completing this work, Dr. Abegglen and his wife returned to Ethiopia to continue their observations on the Cone Rock bands as a longitudinal study. This work meant more to him than any other scientific endeavor. The Abegglens began a series of enclosure experiments in order to test his hypotheses on family formation (fig. 25). The four replicates performed generally agree with the hypothesis, but more would have been required. When, early in 1977, Helga and Jean-Jacques yielded to the mounting political turmoil and had to leave Ethiopia, this became one of his reasons for abandoning science entirely. Jean-Jacques Abegglen died in October 1979. It is sad to think how much he could have added to what he leaves us in this volume.

HANS KUMMER

Acknowledgments

This field study was supported by grant No. 3.363.70 of the Swiss National Science Foundation to Dr. H. Kummer. I wish to thank Dr. Kummer for his guidance and patience during the whole study. I am grateful to Dr. F. Hampel of the University of Zurich for his statistical advice and to H. Sigg and A. Stolba for personal information from their data.

Thanks are due to the ministries of the Ethiopian Government and to the local authorities for their goodwill toward the project. I am indebted to Dr. H. Langenbacher and his collaborators of the Swiss Embassy and to Dr. R. Whipple of the University of Addis Ababa, whose personal efforts on my behalf facilitated the task. I acknowledge the kind hospitality of the inhabitants of Erer, who made our stay in this Ethiopian village a lasting experience.

Finally, I wish to express all my thanks to my wife, Helga. Her persistent efforts and her efficient cooperation during the preparation and realization of the project contributed greatly to the final form of the results.

This manuscript was completed and accepted as a Ph.D. dissertation by the University of Zurich in 1976.

On Socialization
in Hamadryas Baboons

1 Introduction

A. Approach to the Problem

Hamadryas baboons live in one-male units; several one-male units form a band and a few bands occupy a common sleeping site as a troop. This is roughly the social organization of hamadryas baboons according to the field study of Kummer (1968a). The knowledge of this social organization has permitted the continuation of research work on the experimental level (Kummer et al. 1974) by introducing causal aspects that tend to be neglected in field studies (Kummer 1971b). The field study of Kummer has likewise opened the way to the study of social ontogeny. The results that are presented in the following chapters are a step toward an understanding of hamadryas ontogeny. They are based on a field study that was carried out in eastern Ethiopia from March 1971 until July 1972, and they are supplemented by observations from two short field visits in 1973 and 1974, respectively.

What does the study of social ontogeny involve? On the level of a primate society, ontogeny cannot simply mean the development of the ethogram in terms of sex-age classes and behavioral frequencies. A more advanced understanding of primate ontogeny should answer such questions as the following: How does a given social organization influence the development of an individual? What are the possible strategies of a life cycle within the limitations of the social organization? How does the individual life cycle contribute to the regeneration of the social structures? What are the possible modifications of the social organization created by different combinations of individuals? Structurally, an integrated theory of primate ontogeny takes the form of a feedback circuit that relates individual life cycles and social structures to each other. Since primates live in relatively small groups, we may expect that a single generation of individuals cannot realize the full potential of life cycles and of modificatory patterns of social structure. It is then conceivable that the history of a society runs its course as a series of generations that represent varying modificatory patterns of the social organization. Methodically, the descriptive investigation of primate ontogeny requires longitudinal studies, because both the individual and the society are viewed as changing and ongoing phenomena. The longitudinal studies

23

should be complemented by an experimental approach to test hypotheses on the causes of socializing processes.

The multidimensional complexity of primate ontogeny is aptly summarized in the term *socialization*. In a terse formulation it may be said that the study of socialization provides an understanding of society in terms of individual life cycles.

What has been attempted in the investigation of primate socialization to date? A few field studies have provided descriptions of the behavioral development based on a cross-sectional approach: for example, the studies of Baldwin (1969) on squirrel monkeys, of Burton (1972) on barbary monkeys, and of Sugiyama (1965a) on hanuman langurs. These studies have covered all ages of primates, but they have focused on age classes rather than on individuals, and they have largely neglected the causal dimension of socialization.

Much research has been devoted to the early ontogenetical processes, mainly within the mother-infant dyad, under laboratory conditions, for example, the work of Hansen (1966), Harlow and Harlow (1965), and Hinde et al. (1964) on rhesus monkeys; of Rosenblum and Kaufman (1967) on pigtail and bonnet monkeys; and of Rosenblum (1968) on squirrel monkeys. These studies have described the earliest processes of socialization and the effects of social isolation. But they have rarely investigated the precise impact of the earliest processes on later life and have not related their findings to the organization of the free-living society. It is not sufficient to state that the primate infant attains gradual independence from its mother and that through peer contacts it acquires all the necessary skills for its later life as an adult. Such statements are an insufficient link between the laboratory studies on infant behavior and the field studies that concentrate mainly on adult behavior. A review of the literature reveals that only Rosenblum (1971) has attempted to relate the quality of the mother-infant relationship to characteristics of the social organization.

Only a few longitudinal studies on free-ranging societies have been carried out. They comprise the studies of rhesus monkeys on Cayo Santiago, of Japanese monkeys, and of chimpanzees in the Gombe Stream Research Centre. These studies have not emphasized the causes of socializing processes, but they have provided new data on primate behavior that are based on the knowledge of single individuals. They have all demonstrated the importance of genealogical kinship for the cohesion of social subgroups (Koyama 1970, van Lawick-Goodall 1967, Miller et al. 1973, Sade 1965, Yamada 1963) and for the determination of the dominance structure (Kawamura 1965, Koyama 1967, Loy and Loy 1974, Missakian 1972). It has been reported for rhesus monkeys that the mother-dependent rank of females is positively correlated with the reproductive rate of the females and with the survival rate of their infants

(Drickamer 1974). For rhesus monkeys (Missakian 1972, Sade 1968) and Japanese monkeys (Kawamura 1965, Koyama 1967) the course of socialization has revealed a matrilineal pattern of genealogy resulting from the fact that in these species the mother is the only identifiable parent to a young monkey.

To summarize this review, we may conclude that research on socialization has remained rather fragmentary. In an edited book on primate socialization, Poirier (1972) even characterizes socialization as a relatively unknown area of primate behavior. This is amazing if one considers that socialization is undoubtedly a key aspect to a thorough understanding of primate behavior, be it human or nonhuman. Two reasons may have contributed to the discrepancy between the significance of the problem and the relatively little attention spent on it. First, primate behavior is a multidisciplinary field, encompassing zoology, anthropology, and psychology. The recent impetuous development of behavioral studies on primates based on different backgrounds of scientific methodology has produced too many unrelated data and little integrated theory. The lack of such theory discourages investigations of broad phenomena such as socioecology or socialization. Second, the modern structure of scientific research favors the short-term study, which covers an extremely brief section of a society's history but promises a high output of data. Sade (1972) has recently commented on the loss of dimensions that results from the renunciation of longitudinal studies.

What are the contributions to the socialization in hamadryas that the reader will find in the following chapters? This study is, in fact, not the result of a long-term work, since it is mainly based on a cross-sectional approach. However, it prepares the ground for a subsequent longitudinal study. In chapter 2, I shall demonstrate the ecological and social autonomy of the band, which is the basic unit for the socialization processes. Chapter 3 contains behavioral frequencies for all sex-age classes. This conventional set of data represents the quantitative basis for subsequent hypotheses. Chapter 4 attempts to apply concepts derived from experimental studies in triadic differentiation (Kummer et al. 1974) and on bond forming (Kummer 1975) to the behavioral frequencies of immature animals. This step tries to integrate behavioral frequencies into a conceptual framework and to derive causal hypotheses for socialization processes. It remains, however, still an approach in terms of sex-age classes. Chapters 5 and 6 concentrate on family forming, focusing now on particular individuals within one single band. The static concept of the band presented in chapter 1 changes now into one of a dynamic sequence of generations. In chapter 7 the observations on particular individuals will produce evidence on a new subunit of the band that relates the male life cycle to the forming of new families and to the regeneration of the band.

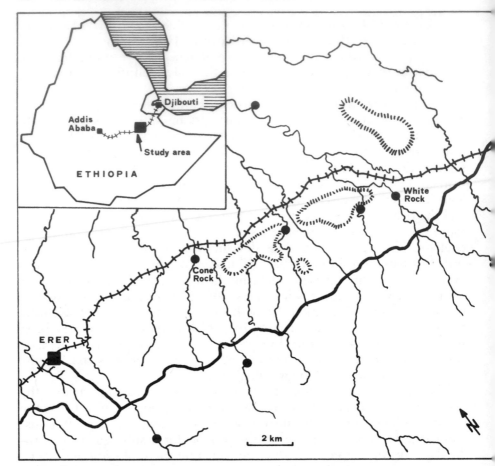

Figure 1. Map of the study area showing the sites of sleeping cliffs (●), the railway line (+++++) and the vehicle track (——). Drawn by H. Sigg and A. Stolba from aerial photographs available at the Institute of Geography, University of Addis Ababa.

The whole framework of hamadryas socialization as it is now viewed, which is presented in chapter 8, was not the theoretical baseline that I hoped to test in the field. Rather, it emerged gradually from the observational interaction with the baboons and from the analytic interaction with the data. The presentation of the results reflects stages of these interactions of subject, observer, and data.

B. Study Site

When Kummer (1968a) surveyed habitats in eastern Ethiopia, he found the region of Erer-Gota to offer representative and favorable conditions to observe hamadryas baboons (fig. 1). A description of the habitat is provided by Kummer (1968a, p. 157). Within the region of Erer-Gota, White Rock and Cone Rock proved to be the most advantageous sites, since they were regularly occupied and both of them were accessible by car. The latter condition was a requirement for transporting traps to the sleeping site of the baboons. Of these two rocks we decided to carry out our prime study at Cone Rock (fig. 2). This choice was influenced by the following conditions. Our preliminary troop counts amounted to about 120 animals at White Rock and 220 animals at Cone Rock. We preferred to study the larger troop, in order to have more individuals per sex-age class in each sample. Moreover, Cone Rock was situated nearer to the village of Erer, which was our living site. This was a practical advantage in terms of time and fuel. A further reason for working at Cone Rock was to diversify our knowledge of hamadryas by choosing a site and troop that was different from the one Kummer (1968a) had examined.

Cone Rock was occupied not only by baboons. In the holes and crevices of the dead volcano lived a population of hyraxes whose species

Figure 2. Cone Rock and its troops viewed from the west. The sleeping sites of Cone Rock troop were located in the upper part of the steep cliff face and on the flat ledges above the cliff. Band II occupied the area on the left side of the rock, band I the area immediately left from the central vertical crevice and band III the area on the right side of the crevice.

Table 1. Sex-Age Classes of Hamadryas Baboons

		Age classes	Characteristics
Immature	Infant	Males N	Hair completely or partially black
		Males B	Hair brown; sitting height about 25 cm
	Juvenile	Males 1	Sitting height about 30 cm
		Males 2	Sitting height about 40 cm
		Males 3	Longer hair at the sides of the head; sitting height about 48 cm
Mature	Subadult	Males M−	First signs of mantle; sitting height about 50 cm
		Males M±	Mantle half developed; sitting height about 55 cm
		Males M+	Mantle fully developed, but browner than adult mantle; hair at the sides of the head not fully developed; sitting height about 60 cm
	Adult	Males a	Mantle and hair at the sides of the head fully developed; sitting height about 65 cm
Immature	Infant	Females N	Hair completely or partially black
		Females B	Hair brown; sitting height about 25 cm
	Juvenile	Females 1	Sitting height about 30 cm
		Females 2	Sitting height about 38 cm
		Females 3	Sitting height about 45 cm
	Adult	Females a	Sitting height about 50-55 cm

could not be determined. Hyraxes and baboons interacted rarely, since the hyraxes appeared at the surface of the rock during the absence of the baboons. A few times, however, juveniles and subadult males threatened hyraxes that had come out of their holes. Once a subadult male was bitten on the tail by a hyrax, and once a subadult male carried a dead hyrax (which he probably had killed) in his mouth. For more than one week, a pair of Ruppell's Griffon vultures *(Gyps ruppellii)* attempted to nest on a cliff ledge that was a sleeping site of hamadryas. A similar attempt was made by Egyptian vultures *(Neophron percnopterus)*. The nesting attempts were unsuccessful because the birds, which were not bothered during the day, were threatened and chased by the baboons in the evening. The vultures finally left the rock.

To reach the sleeping site of the baboons and to transport the traps we used a short-wheelbase Toyota land-cruiser. Observations were made with 7 × 50 Kern binoculars and behavior was recorded on portable cassette recorders.

C. Sex-Age Classes

With some minor modifications I have used the definitions of sex-age classes given by Kummer (1968a). Our classes are summarized in Table 1. In comparison with Table 1 of Kummer (1968a, p. 9), the following differences should be noted:

Kummer's class of the one-year-old animals has been split up into two classes, labeled animals B and 1 respectively. The age class B is composed of infants between 6 and 12 months old; age class 1 includes Kummer's 1- to 1.5-year-old animals. Our infants N and B correspond with the infants 1 and 2 in anubis baboons (Ransom and Rowell 1972) and in cynocephalus baboons (Altmann and Altmann 1970).

The experiences from our field observations indicate that the estimates of age as given by Kummer (1968a) are probably too low (see figs. 26 and 27). While using the same classification as Kummer, I shall apply the terms of animals 1, 2, and 3 without prejudging any age statements.

The distinction of subadult and adult females has been dropped, because no use is made of it. Any female older than a female 3 is referred to as an adult female.

D. Trapping and Marking

To capture baboons we used four traps, consisting of an iron frame and wire net walls, and measuring some 1.2 meters in length and 0.7 ×

0.7 meters on the entrance side. A sliding door at the narrow side could be released by means of a string. For bait we always used corn. Trapping took place generally in the morning, so that we could anaesthetize and mark the animals during the day and release them on the same evening. Continuous trapping proved very disturbing to the troop; therefore, we trapped only once per week or per fortnight. The mere chance of who entered the traps determined the sample of marked individuals. We had no other criteria for selecting individuals, because we did not know the band and family membership of the animals at that time.

During the dry season, anaesthetizing and marking could be done at Cone Rock after the troop's departure. In the rainy season enough water and food was available so that the troop could afford to stay near the sleeping cliff and to guard the traps. During the rainy period we approached the traps by car and took the captured animals to our house for the marking procedure.

For an anaesthetic we used Parkesernyl® (phencyclidinhydrochloride), which was injected into the muscles of the thigh or the arm. The dose was 2 milligrams Parkesernyl per kilogram of body weight. The narcosis lasted 85 minutes on the average. In general the animals showed no postnarcotic effects, and all could be released within twelve hours in normal condition.

At the beginning we used Polycolor® and Nyanzole[1] to dye the hair. In subadult males the resulting black stain remained visible for a few months and by then the male was identifiable by his morphological features. In juveniles the dyes proved to be unsatisfactory, because the stains disappeared within two or three weeks. Marking of juveniles became possible when we applied colored plastic ear tags (Rotag®). The tags did not seem to bother the animals, and in the troop they were soon ignored by the other baboons.

E. Samples

Our main study period was from May 1971 until July 1972, comprising somewhat more than 500 observation hours. In addition, two short field visits were made in April/May 1973 with about 70 observation hours, and in February 1974, also with some 70 hours. The following sets of data were collected:

Close-study data: We sampled and marked 39 individuals of different sex-age classes, excluding harem leaders and adult females (tab. 2). Seven-

1. I would like to thank the Profar S. A. of Geneva, Switzerland, which provided us generously with Polycolor hair dye. I also thank Mr. T. Martin of the Delta Primate Research Center in Covington, La., U.S.A., who provided us with the Nyanzole.

Table 2.   Sex–Age Classes of the Close–Study Animals

Age class	N	B	1	2	3	M–	M+	M+	a	Total
Males	4	4 (1)	4 (0)	4 (3)	4 (0)	6 (3)	4 (1)	4 (2)	4 (2)	26
Females	5	4 (2)	3 (2)	4 (0)	4 (3)	–	–	–	–	13

The figures give the number of individuals per sex–age class which were marked. Figures in parentheses indicate the number of individuals which entered the respective age class from the lower one during the study period.

teen of these were members of Band I. These individuals were observed during the morning and evening periods at Cone Rock. Behavior was sampled in time units with a timer dividing the records into 15-second intervals. The duration of observation on any one individual varied according to conditions of visibility. Seventy-five percent of the observation sessions were within 10 and 25 minutes (minimum 4 minutes, maximum 80 minutes). Neither the beginning nor the end of the record was determined by the animal's behavior. Before driving to Cone Rock, we decided which animals were to be observed, and as soon as the respective individuals were detected in the troop we started recording. Per visit at the rock, we could observe one to four individuals. We attempted to observe the individuals in a regular cycle. The total of the time-unit samples on marked individuals will be referred to as the *close-study data.* They comprise a total observation time of 196.5 hours. The quantitative records were supplemented by semiquantitative short protocols and by qualitative notes.

Broad-sample data: Another set of data was collected on all three bands in the following way. We combined the age classes B and 1 into one single class and did the same for age classes M± and M+. In the age class of the adult males we included all adult males, whereas the close study excluded the harem leaders. In addition we included the age class of the adult females, which were neglected in the close study. During 80 observation sessions between October 1971 and July 1972 we took quantitative records of one minute per individual for all visible animals of a particular sex-age class. This procedure was repeated until we had collected 60 one-minute samples for each sex-age class. The same sample of 60 one-minute records per class was also collected at White Rock during 25 observation sessions between October 1971 and July 1972. These time-unit samples on all animals will be referred to as *broad-sample data*. They comprise a total observation time of 24 hours.

Nonrandom samples: On a few occasions we took records for individuals (marked and unmarked) that displayed a behavior of special interest. These nonrandom observations represent a total observation time of 11.5 hours.

Notifying-event samples: The behavior of subadult males directed our attention toward notifying, which is a ritualized presenting among mature males (tab. 9). Notifying occurred rarely in the time-unit samples of the close study. Therefore we decided to register each occurrence of notifying that involved mature males of our prime study band (band I at Cone Rock). The result was 468 notifying events.

   At the end of our study, the animals had become so habituated to our presence that we could follow the traveling troop at close distance. We followed band I fifteen times for the first hour of the foraging march and recorded all notifying events (sample size = 132). Moreover, we described in qualitative and semiquantitative terms several aspects of coordination of travel.

Band counts: The counting of a band was easiest in the evening, when the bands reached the sleeping cliff as separate units and when the animals marched more or less in a line. Between September 1971 and April 1972 we counted the numbers of the three bands 21, 19, and 16 times, respectively.

Band censuses: Independently from the counts, we determined the band compositions in terms of sex-age classes. These censuses were possible only under excellent conditions of observation. They were carried out for bands II and III only, because a more precise census would be available for the prime study band I. In each of the two bands, seven censuses were taken between February and May 1972.

Band leaving and arrival: From October 1971 until July 1972 we noted for each band the time and direction of departure in the morning and of arrival in the evening. We registered 132 departures and 130 arrivals for each band.

Band I—data: Band I of Cone Rock became our prime study band. In this band we knew all the mature males individually. Therefore, we could determine the precise composition of all one-male units. This refined census was repeated in intervals of two months and again during the short field visits in 1973 and 1974.

2   The Band Unit

The ontogeny of each individual takes place within the framework of particular social units. In the course of our work we gained the impression that the band represents a basic unit for the individual life cycle. The band has been described by Kummer (1968a), but until now it has lacked a quantitative description and evidence of its social functions. The aim of this chapter is to provide numerical data on the band and to demonstrate that it is the relatively closed frame within which the life cycles of its members take place.

A. The Bands of Cone Rock

The troop of Cone Rock consisted of three bands that occupied the main rock during our field stay. On the other side of the dry riverbed, opposite Cone Rock, one more band usually spent the night on a separate cliff. This fourth band was not studied, since it normally was outside of our visual range and, moreover, it occupied its sleeping site irregularly. The three bands on the main rock, however, were present whenever we visited the rock in 1971/72.

Soon we learned to identify the bands by means of the marked individuals. We referred to them as bands I, II, and III. On the rock the three bands occupied sleeping areas with stable boundaries. These areas remained the same even in times of social tension and change. The stable sleeping areas do not imply that the bands were restricted at all times to their own area or that they defended them. Usually, the band that arrived first in the evening occupied large parts of Cone Rock. The following bands then joined the first one without provoking any antagonism. In the course of the evening the one-male units moved gradually so that before dark all bands were within their sleeping areas. When the last band arrived at the end of dusk, its settlement on the rock was accompanied by intensive contact vocalization throughout the whole troop. In the morning the bands soon left their places and spread out over the rock. The stable sleeping area, therefore, became visible only in the morning and the evening between darkness and the social period at the rock.

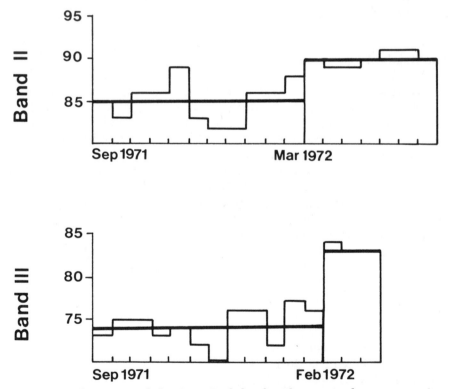

Figure 3. The counts of the Cone Rock bands. The repeated counts are given in chronological order. In spring 1972 the size of the bands has increased due to the birth of new infants. The thick line represents the means of the counts before and after the birth season. Points on the abscissa represent counts actually taken at irregular intervals.

The results of the band counts are graphically summarized in figure 3. The differences between the single counts were probably due to errors in counting. Table 3 contains the mean sex-age compositions of the bands for March 1972. The repeated censuses of band I are summarized in table 4. The composition of band I remained fairly constant. In particular, no one-male unit of band I was ever observed to change band. The same holds true of the twelve one-male units that we had identified in bands II and III. Table 5 represents the individuals that increased the size of band I between May 1971 and February 1974. Twenty births and 3 immigrations were recorded during this period. Possibly a few more births occurred after August 1972, which were not noticed if the infants died before my field visits in April 1973 and February 1974. Of the 20 newborn infants 5 were males and 15 were females. Between February and September 1974, 10 more births were recorded—3 males and 7 females (Sigg and Stolba, personal communication). In band I male births occurred less frequently than female (sex ratio 1 : 2.75 in 3.5 years). This result holds true only of band I and is not supported by the figures for bands II and III (see tab. 3) nor by the findings of Kummer (1968a).

Between May 1971 and February 1974 only three individuals entered band I—two adult males and one adult female. The scars of the two males indicated that they had lost their females by fight (see section 6c). One of them was an old male, who probably came from band IV and joined band I in June 1972. He was observed alternately in bands I and IV until he disappeared at the end of our main study (see tab. 6). We assume that he died. The other male was younger; he came from band III and entered band I in February 1974, where he was still found in September 1974 (Sigg and Stolba personal communication). The adult female was taken over by a band I leader from an unknown troop. She was lost again after a few days (see tab. 6). Between August 1972 and April 1973 the largest one-male unit of band I was dissolved (see fig. 22). The breakup produced numerous changes of group memberships and many losses of individuals (tab. 6).

Table 6 summarizes the losses in band I. Unfortunately, we do not know what happened to the animals that disappeared. Five of the 7 males and 7 of the 10 females disappeared between the end of our main study and the field visit in April 1973. During the breakup of the largest one-male unit of band I, some of the females were possibly taken over by leaders of other bands; we cannot verify this possibility, because none of the females was marked. Three of the 7 males were marked, but none of them was ever sighted again in any other band around Cone Rock. One of them was an old male who had lost his females in 1972 (see fig. 21). Another was a younger leader, who probably became involved in fights over females during the breakup of the large one-male unit; he might

Table 3. Sex-Age Composition of the Cone Rock Troop in March 1972

	Males							Females						Total
	a	sa	3	2	1+B	N	Total	a	3	2	1+B	N	Total	
Band I	8	5	0	4	5	2	24	19	1	4	13	1	38	62
Band II	10	5	2	5	12	4	38	29	5	4	9	5	52	90
Band III	9	9	2	5	11	4	40	29	1	2	7	3	42	82
Cone Rock Troop	27	19	4	14	28	10	102	77	7	10	29	9	132	234
White Rock Right Band	12	2	5	4	5	3	31	33	2	6	5	3	49	80

The figures on the White Rock band refer to the situation in 1968 and were collected by Kummer (unpublished data).

Table 4. Censuses on Band I

	Males							Females						Total
	a	sa	3	2	1+B	N	Total	a	3	2	1+B	N	Total	
August 1971	7	6	0	3	6	1	23	19	0	2	4	13	38	61
October 1971	7	6	0	3	5	1	22	19	1	2	8	8	38	60
December 1971	7	6	0	3	5	1	22	19	1	3	11	4	38	60
February 1972	8	5	0	3	6	0	22	19	1	4	11	3	38	60
April 1972	8	5	0	4	5	2	24	19	1	4	13	1	38	62
June 1972	8	5	0	4	5	3	25	19	2	4	11	2	38	63
August 1972	8	5	0	4	5	4	26	19	2	4	11	2	38	64
April 1973	7	4	2	3	5	0	21	15	3	2	10	1	31	52
February 1974	9	4	1	4	4	0	22	18	1	5	7	4	35	57

Table 5. Gains (Immigration and Birth) of Individuals between May 1971 and February 1974 in Band I

		Males				Females			Total
		a	sa	B-3	N	a	B-3	N	
May August	1971–1972	1	0	0	5	1	0	11	18
September April	1972–1973	0	0	0	0	0	0	0	0
May February	1973–1974	1	0	0	0	0	0	4	5
Total		2	0	0	5	1	0	15	23

Table 6. Losses (Emigration or Death) of Individuals between May 1971 and February 1974 in Band I

		Males				Females			Total
		a	sa	B-3	N	a	B-3	N	
May August	1971–1972	1	0	1	0	1	1	1	5
September April	1972–1973	2	0	3	0	5	2	0	12
May February	1973–1974	0	0	0	0	0	0	0	0
Total		3	0	4	0	6	3	1	17

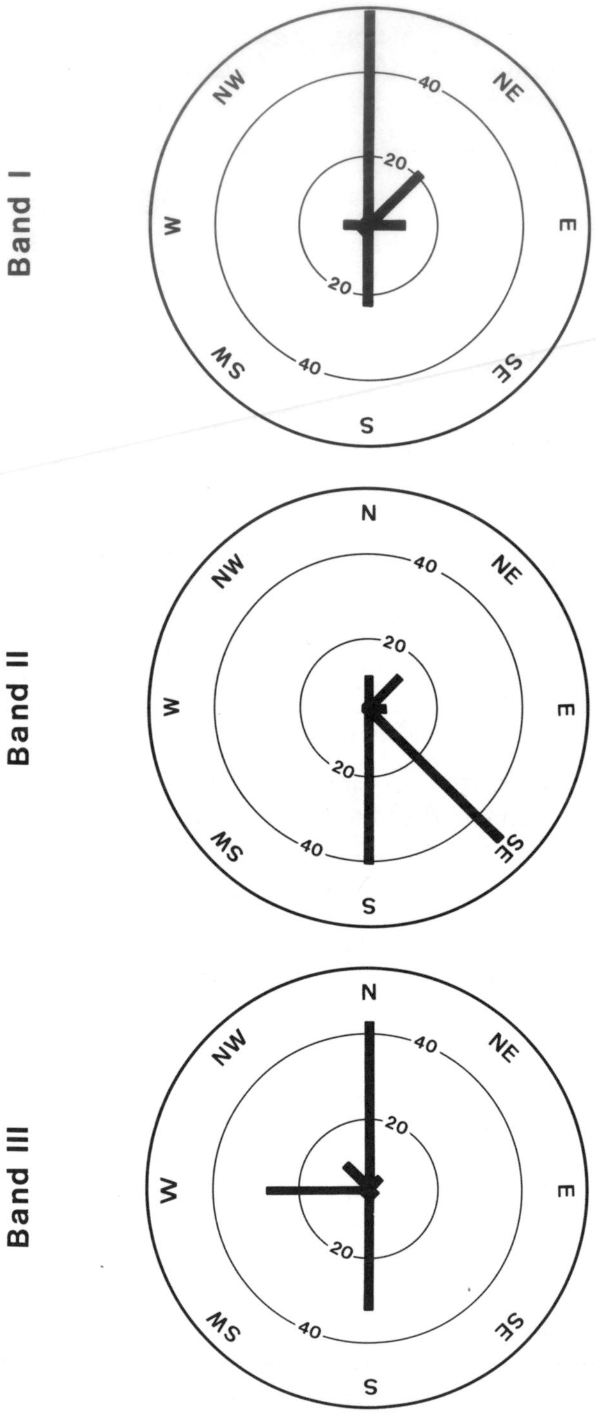

Figure 4. The directions of arrival at Cone Rock in the evening. The figure gives the frequency of the directions which the Cone Rock bands used within the last 200–300 meters in 132 evenings between October 1971 and July 1972.

have been severely wounded because he already had a handicap of a damaged eye. The third male belonged to the one-male unit that was dissolved. We assume that these three males died. Of the 40 changes in band size given in tables 5 and 6, 20 were due to birth and 20 were related to the breakup of one-male units.

According to our figures on the Cone Rock bands, the band appears as a stable unit that probably permits a long-term identification. Some 25 visits at White Rock between August 1971 and July 1972 qualitatively revealed the same characteristics for the bands there. We found two bands with about 40 and 70 animals. They occupied separate parts of the rock, with the smaller band on the left and the larger band on the right side of White Rock.

B. The Band as an Ecological Unit

Kummer (1971a) gives the impression that after the departure from the sleeping cliff the troop divides into bands that represent ecological units of foraging. To get a more detailed picture of this aspect we have recorded times and directions of departures and arrivals of the bands from and at Cone Rock. We turn first to the more informative *arrival* of the bands.

Figure 4 shows how many times each band arrived at Cone Rock from the different directions, each recorded as the average direction for the last 200–300 meters. The figure demonstrates that each band arrived from preferential directions. Band I, in nearly three-quarters of the 132 evenings, came from the northern to eastern sectors. In about 80% of the evenings band II arrived from the south to southeast, band III arrived from the western direction more often than the other bands did. For band III the southwestern arrivals tend to be underrepresented in favor of the southern direction, due to the nature of the terrain: when the band approached Cone Rock from southwest it had to make a detour to pass around a steep cliff. As a consequence the band reached the dry riverbed at some hundred meters south of Cone Rock. The monthly means of the arrival times are given in figure 5. The time of arrival was recorded as the moment when the first animal of a band crossed the defined boundary of the sleeping cliff. Figure 5 suggests a constant sequence of bands in reaching Cone Rock. In 90% of the arrivals, band I was the first band at the rock; band II tended to be second and band III last to return from the foraging march. Each band arrived as a separate unit at Cone Rock both in terms of direction (tab. 7) and time of arrival (tab. 8). The band-specific directions of arrival suggest that each band used a different foraging area.

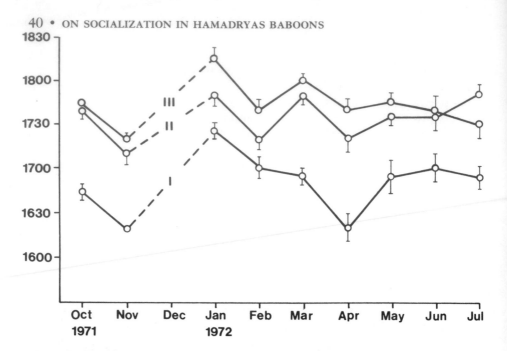

Figure 5. Monthly means and standard deviations of mean for the arrival time of the three Cone Rock bands. The figures are based on a total of 135 arrivals for each band. The Roman numbers refer to the bands.

The time of *departure* was defined as the moment when the last animals of a band climbed down the cliff for the foraging march. Directions of departure differ less among the bands than directions of arrival (see fig. 6). While the baboons arrived in separate bands, they tended to leave at the same time and in the same direction, as a troop (tab. 7 and 8). Figure 6 shows that the troop did not take all the directions with equal frequencies. In 44% of the 130 mornings the troop left to the north or to the south; that is, it followed the dry riverbed of Cone Rock, which runs from south to north. In 40% the troop took the sectors from northeast to southeast, which led near a range of small hills with richer vegetation. The common departure of the troop indicates social facilitation and attraction among the bands. This assumption is supported by two other observations: (1) band IV, which slept opposite to Cone Rock on a separate cliff, regularly joined Cone Rock troop for the departure; (2) occasionally, a strange band from an unknown rock passed the night on Cone Rock. In the following mornings the alien band always left very early. The Cone Rock troop, which usually would have stayed longer, would follow the strange band at the same time and in the same direction.

What direction did a band take when it did not leave with any of the other bands? In 14 of 19 cases band I left for the sector to the northeast;

in 16 of 23 cases band II went to south or southeast; and in 23 of 29 cases band III departed toward the northwest. These directions coincide with the most frequent directions of arrival (fig. 4). This means that when the bands did not leave as a troop, they would take their particular direction, which probably brought them more directly to their foraging ranges.

Figure 7 gives the monthly means of the departure times for Cone Rock troop. The time of leaving seems roughly correlated with the astronomic sunrise, as Kummer (1968a) had suggested. The departures of November 1971 occurred earlier than one would expect on this assumption. A possible explanation is that in this month a one-male unit of band II was divided up (see chapter 6), and that the resulting social tension led the bands to separate as early as possible in the morning. An early departure in uncertain situations was also observed when alien bands passed the night on Cone Rock (see below and Kummer 1968a).

Figure 8 summarizes the monthly means of the *duration* of the foraging march. It suggests that the foraging march tends to be shorter during the

Table 7. Differentiation of Directions of Arrival and Departure among Cone Rock Bands

			Arrival	Departure
All bands arrive or leave in the same direction			8	81
Band I	arrives or leaves in a different direction than the other bands		6	8
Band II			40	12
Band III			13	18
All bands arrive or leave in different directions			65	11
Total number of cases			132	130

Table 8. Differentiation of Time of Arrival and Departure among Cone Rock Bands

			Arrival	Departure
All bands arrive or leave at the same time			2	108
Band I	arrives or leaves at a different time than the other bands		22	2
Band II			4	5
Band III			3	12
All bands arrive or leave at different times			101	3
Total number of cases			132	130

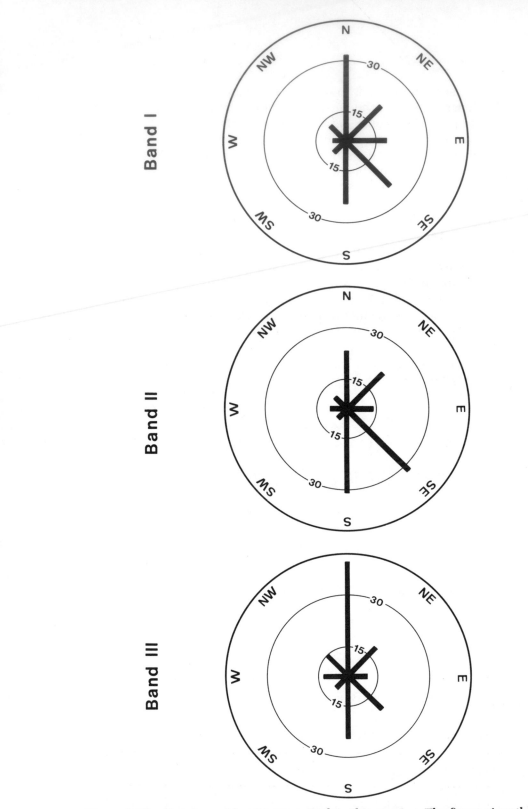

Figure 6. The directions of leaving Cone Rock in the morning. The figure gives the frequency of the directions which the Cone Rock bands took for the first 200–300 meters in 130 departures between October 1971 and July 1972.

rainy season. The fact that the mean duration of travel did not decrease in July 1972 in spite of the rains might again be explained by the breakup of a one-male unit, which took place in band I during this period. In the morning the bands departed very early in accordance with the early sunrise (fig. 7). In the evening the breakup of the one-male unit led to interband avoidance and therefore to a late settling of the bands on Cone Rock.

Figures 7 and 8 suggest the following hypotheses: the time of sunrise decides the normal time of departure; the availability of water and food, which is regulated by the amount of precipitation, determines the duration of the foraging march. The times of departure and arrival can be affected by social disturbances that lead the bands to avoid one another.

Our data on departure and arrival of the baboons emphasize the autonomy of the band as a foraging unit. Again, the Cone Rock data are qualitatively supported by our observations at White Rock: the two bands of White Rock troop tended to leave at the same time and in the same direction, and to arrive as separate units. The separate arrival was also observed by Kummer (1968a). We had the qualitative impression that the larger White Rock band arrived frequently from the northern to

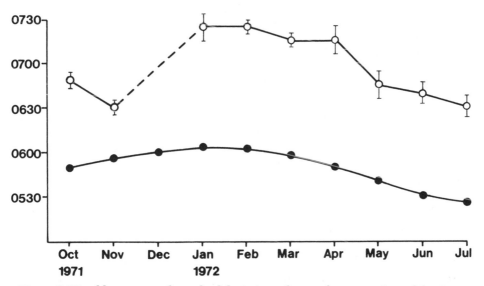

Figure 7. Monthly means and standard deviations of mean departure time of the Cone Rock troop. The figures are based on a total of 133 departures. The lower curve gives the monthly means of the astronomic sunrise, whose calculation I owe to Mr. H. J. Pfister from the Swiss Federal Observatory, Zurich (Switzerland). The monthly mean departure time is positively correlated with the time of sunrise (Spearman rank correlation coefficient 0.83, p = 0.01).

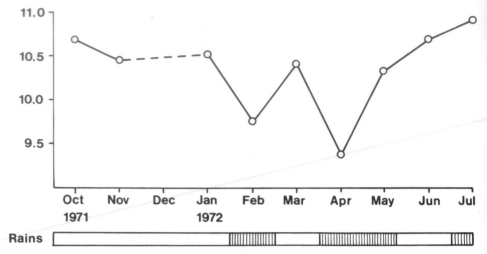

Figure 8. Monthly means of the duration of the foraging march. The standard deviations of means are not presented, because they have the order of magnitude of a few minutes only. The curve suggests that the foraging march tends to be shorter during some of the rainy months.

southeastern sector, whereas the smaller band returned frequently from the southwestern to northwestern sector.

C. The Band as a Social Unit

1. Troop Cohesion and Interband Encounters

Kummer (1968a) reported that bands may use several sleeping cliffs in turn and that interband encounters may range from antagonism to tolerance. This is supported by our own observations. During our main study period in 1971/72 the three bands of Cone Rock passed each night together on the main rock. We registered only three agonistic incidents in Cone Rock troop that involved short fights between males of different bands. These incidents occurred in a period of structural changes within the bands (see chapter 6).

Band IV, which occupied its own sleeping cliff opposite Cone Rock, was sometimes absent for several weeks. Four times a strange band of 60 to 70 animals arrived at Cone Rock and passed the night there. Probably it was the same band each time because it always contained an anubis female that originated from the transplantation experiments of Kummer et al. (1970). The arrival of this band did not produce any antagonism. Five times other alien bands attempted to approach Cone Rock. In one of these cases the alien band responded to the alarm barks of Cone Rock troop with an immediate retreat. In the four other cases the encounter ended in a fight with Cone Rock troop. The fights occurred in the phases that Kummer (1968a) has described: contraction of each band; fights that took place mostly as duels between mature males; finally, separation between the bands. Twice the alien bands succeeded in settling on Cone Rock after darkness had fallen; once the alien band remained on a rocky ridge near Cone Rock; and once it occupied the empty sleeping site of band IV. The next morning, after all these encounters, Cone Rock was already empty at dawn.

During my visit in April 1973, band II had left Cone Rock and slept somewhere in the hill range southeast from Cone Rock, where I found the band very early one morning. In February 1974 only band I was regularly on Cone Rock. Band II occupied a sleeping cliff one kilometer south of Cone Rock, and band III was found on a small sleeping cliff about three kilometers northwest of Cone Rock. It is worth noting that the new cliffs of bands II and III are situated in the most frequent directions of arrival during 1971/72. Band IV still occupied its site opposite Cone Rock. I do not know why bands II and III left Cone Rock. The new sleeping sites of these two bands appeared of inferior quality compared with Cone Rock. A possible explanation for the separation of the bands is the effect of the drought in Ethiopia, which led to an increased competition for the scarce water resources between the bands and between baboons and human nomads.

Our observations show that, although each band represented an autonomous unit, there was some cohesion on the troop level. Between May 1971 and February 1974, bands I to IV have regularly been found in the region around Cone Rock.

2. Interactions of Juveniles across Band Units

During the social periods in the morning and evening, the three bands were spatially segregated, and intermingling was rare. From qualitative evidence we estimate that more than 90% of the play groups contained juveniles of the same band. Our close study and short protocols comprise nine instances of short interband encounters that involved juvenile animals. Six of them were play interactions among males. In one of the six interactions, play was followed by mounting. One further interaction was play between a juvenile male and a juvenile female. Grooming interactions of short duration were seen twice: once a male 1 groomed a male 3 and on another occasion a male 2 groomed a female B. Interband play seemed most likely when the bands were ready to leave in the morning while deciding on the direction of travel. In this situation the bands were sitting close together on an edge of the rock and the juveniles of different bands could easily meet. Another opportunity for interband play occurred when the bands were gradually shifting on the rock, during morning and evening. Then the individuals of one band, which had occupied a play site, were joined by members of the substituting band. For a few minutes an interband play group resulted, until the juveniles of the first band left the spot and the play site was occupied only by animals of the second band.

Besides these short interactions a few cases of longer interactions were registered:

One evening Shadow, a subadult male of band III, arrived on the rock about half an hour before the rest of his band. The other bands were already present. Shadow was groomed by Mango, a male 2 of band I. After some minutes a conflict between two other males caused the grooming pair to move. Mango returned to his band and Shadow was approached by Iaso, a male 3 from band II. Iaso groomed Shadow until band III arrived and Steward, who was the usual grooming partner of Shadow, appeared. Steward and Shadow embraced each other and started grooming. Iaso remained nearby, fiddling the ground, then went back to his band.

10. 5. 72: Band I is in the dry riverbed in front of Cone Rock. Mango, a male 2, sits in his one-male unit and is groomed by his mother. When the band moves slowly to the rock, Mango follows, then suddenly leaves his band together with another male 2. The two males climb back to the riverbed, cross it, and climb slowly to the sleeping site of band IV, which has already arrived there. Mango approaches a solitary young adult male and grooms him for about two minutes. The other male 2 remains sitting nearby. Then Mango attempts to move farther on but is threatened by a young subadult male. Immediately another subadult male lunges at the aggressor, while Mango and his companion pass quickly. Mango approaches a second solitary young adult male and grooms him for about five minutes. Again the other male 2 remains sitting nearby. Then both juveniles settle in a niche on the cliff of band IV and remain there until dark.

15. 5. 72: We have not seen Mango until this day. In the morning we find him still in band IV, which he follows for the foraging march. The male 2 that had accompanied Mango is back in band I.

16. 5. 72: In the evening we observe Mango again in band IV, where he plays intensely with another juvenile male.

23. 5. 72: In the morning we find Mango back in band I. After this date he remained in his band until we left Erer in August 1972. When I visited Cone Rock in April 1973, Mango's parental unit had been dissolved (fig. 22) and Mango was not in the band anymore. I never saw him again in any band.

Such a temporary interband visit occurred again in February 1974. Bub, a young subadult male, and Zabli, a male 2 from the same one-male unit, left band I and stayed for about one week in band III. In this case the visit may have been so extended because, after the transfer of Bub and Zabli, bands I and III occupied separate sleeping cliffs for about one week, so that the males probably could not move back to their own band.

Until now there has been no evidence of a juvenile's remaining in another band for more than a few days. The observed instances of interband interactions suggest that juvenile males interact and visit with other bands more often than juvenile females.

3. Interactions of Mature Males across Band Units

Among older animals, only males were seen to interact across band units. These interactions consisted usually of notifying, which is analyzed in chapter 7. The two adult males that immigrated to band I in 1973 and 1974 (tab. 5), respectively, were frequently seen grooming with the subadult male Stupsie. In chapter 7 I shall give a possible explanation of why Stupsie became a favorite partner for these immigrant males. None of the immigrant males was ever seen to interact with females of band I.

4. Interactions of Infants across Band Units

The interband encounters that involved infants N and B revealed a tendency to prevent interactions across band units. We have recorded nine instances. I shall give four examples as illustrations:

Prima, a female N of band I, has climbed a few meters down a cliff and stays near a one-male unit in band III. The leader of Prima's one-male unit gets up, places himself in a position above Prima, and stares down at the infant. Prima runs back to her maternal unit with staccato-coughing vocalization.

Resola, another female N of band I, approaches a male infant of band III. A male 1 of Resola's one-male unit rushes at her and grasps her for a few seconds. Resola tries a few more times to make contact with the male infant, but each time a male 1 and a female B from her band interfere. They grab Resola or place themselves between the infants and prevent any interaction.

After arriving at Cone Rock, some juveniles of band III form a play group on the band's periphery toward band I. Resola approaches the group and intends to play with Caramel, a male infant N of band III. Zabli, a male 1 of Resola's one-male unit, wants to interfere, but a male 2 from band III, who is a favorite play partner of Caramel, lunges at Zabli. Immediately, Zabli's mother rushes at the male 2 and threatens him. The male 2 threatens back while grabbing Caramel. He is supported by a subadult male of band III, who sits nearby and now also threatens Zabli's mother. After a few seconds the animals calm down, but no further interaction across the band limit occurs.

In front of Cone Rock, bands I and II have settled in the dry riverbed. A play group has brought together juveniles from both bands. Suddenly screaming arises, and a male 1 of the band I bites a female B of band II. There is loud screaming and Stumpe, a dominant male of band II, rushes into the play group. The playing animals disperse quickly. Within a few seconds the excitement spreads and the two bands draw up against one another, while the males utter "bahu" barks and "oohu" roars. Then band I runs away in closed formation and settles again some 30 meters away.

These records make it evident that active efforts are undertaken to suppress interband encounters of infants. We never observed a leader interfere personally to bring back an infant within the band, unless the infant was threatened by a subadult male. Screaming and chasing in intraband play groups were common but never caused a dominant male to intervene.

D. Summary and Discussion

The observations on the Cone Rock bands demonstrate that the band is a stable and autonomous unit: (1) the band remains spatially segregated from other bands and occupies a separate sleeping area on the rock; (2) its sex-age and harem-unit composition is fairly stable over time; (3) it represents an autonomous unit of foraging and probably uses a preferential foraging area.

Brief interactions among members of different bands and temporary transfers to other bands occur. Attempts of infants to interact across band units are suppressed. In hamadryas as well as in hanuman langurs (Sugiyama 1965a), intergroup fights are risked when infants enter another troop. Interband encounters of juveniles are not frequent, but there is no evidence that such interactions are suppressed. We assume that infants do not yet know their own band membership nor that of other animals; therefore they have to be prevented from entering other bands in order to avoid interband fights and, possibly, to facilitate the smooth separation of the bands in the morning. It appears advantageous that mainly juveniles hinder the engagement of inexperienced infants in interband encounters: the infant's mother is often not able to intervene since the leader does not permit her to leave him; moreover, it seems inefficient to involve the leader's intervention and thus risk an interband fight each time an infant enters another band. Juveniles apparently know about band membership; they are mobile and provoke little antagonism.

They are therefore the best qualified to prevent interband encounters of infants.

Besides temporary interactions between individuals of different bands, we have found very few permanent transfers to other bands. At present it seems that leaving a band is not common in hamadryas. Adult males that have lost their females and females that are taken over by leaders of other bands are the most likely to change their band.

Hamadryas males change bands less often than males of other baboon species change their groups. In three years, only two adult males entered band I of Cone Rock troop. Of the 8 males that disappeared in band I, none was ever found in another band. Rowell (1966, 1969) reports for a population of anubis baboons that the number of large juvenile and adult males could change daily, and that some of the males must have transferred almost every time two troops approached each other. Adult anubis females were never seen to move out of their group. In the cynocephalus baboons of Amboseli, Altmann and Altmann (1970) observed three immigrations and three emigrations of adult males in a troop of about 40 animals during a period of 10 months. The authors report that cynocephalus males may leave the group a few days after having lost a fight. This parallels our observation of the two adult males that moved to band I after having lost their females in fights. In contrast to our males, the cynocephalus males that changed their group were seen to copulate with several females in the new group. In chacma baboons, males can live temporarily or permanently apart from their troops, whereas females are anxious not to lose contact with the troop (Stoltz 1972); however, chacma females are accepted in other troops when experimentally transferred (Saayman 1968), just as hamadryas females are also accepted in other troops (Kummer 1968a).

Among other Cercopithecoidea, migrating or solitary males, but not females, have been observed in rhesus monkeys (Koford 1966, Lindburg 1969), in Japanese monkeys (Kawanaka 1973, Nishida 1966), in vervets (Gartlan and Brain 1968, Struhsaker 1967), in hanuman langurs (Sugiyama 1967), and in Nilgiri langurs (Poirier 1969). Intergroup mating during encounters of different troops has been reported for rhesus monkeys (Boelkins and Wilson 1972, Hausfater 1972, Lindburg 1971). In contrast to these species, the hamadryas males that temporarily leave their band have hardly any chance to mate in another band unless they acquire females there. Even if a male happens to copulate with a female unnoticed by her leader, the reproductive effect remains minimal, since baboons require a series of mountings to achieve ejaculation (Hall and DeVore 1965, Kummer 1968a). We assume, therefore, that the interactions across band units and the temporary transfers of hamadryas males to other bands contribute very little to the genetic exchange between bands.

On the other hand, the social organization of hamadryas offers the possibility of female transfers because such transfers can be forcefully imposed by males of other bands. Observations at the species border of anubis and hamadryas baboons suggest that hamadryas males take over females from anubis troops (Kummer et al. 1970, Nagel 1973). Moreover, according to the observations of Kummer (1968a), which are supported by our own, fights between bands consist of—or result in— fighting over females. Two attempts to take over a female from another band are reported in section 6A. The data presently available suggest that genetic exchange between bands involves females rather than males in hamadryas baboons. If this is confirmed, hamadryas differ from previously studied Cercopithecoidea. Other instances of female transfer are known only in chimpanzees (Nishida and Kawanaka 1972).

The ecological autonomy of the bands, which seem to use preferential foraging areas, supports the hypothesis of Kummer (1968a, 1971a) that the multilevel organization of hamadryas represents an adaptation to diverging needs: the scarcity of sleeping cliffs requires large concentrations; the sparseness of food favors dispersed social units. If band-specific foraging ranges were confirmed, the hamadryas band would not differ so markedly from troops of Japanese monkeys as Kawanaka (1973) suggests. The only difference between what Kawanaka refers to as a local concentration of troops and the hamadryas troop consists in the fact that the hamadryas bands occupy a common sleeping site.

3   Frequencies of Behavior

A. Terms of Behavioral Classes and Frequencies

I have used the terms of behavioral classes that Kummer (1968a, p. 180) listed in his monograph. For quantitative analysis, elements of behavior have been grouped into categories. These categories are defined in table 9. No regard is paid to who is performing the activity or what function the behavior serves. The term *maternal behavior*, therefore, is also applied when it is performed by juveniles or by subadult males. The term *sexual behavior* is also used when it occurs among juveniles or among males.

Interobserver reliability was calculated as

$$ R = \frac{C}{C + O + D} $$

where C is the number of correct, O the number of omitted, and D the number of differing elements in one or the other record. We started to collect data after having reached a rate of 0.85.

In the close study the records of each individual were pooled over time. Since the totals of observation time differed among the individuals, frequencies are expressed as percentages of 15-second intervals,[1] including at least one occurrence of the behavior. When an individual changed age class during our study period, the frequencies were calculated separately for the two age classes.

The broad sample consists of quasi-independent, individual records of one minute. The number of records was the same for all sex-age classes, and frequencies are expressed as the number of one-minute records containing at least one occurrence of the behavior.

B. Statistics

The data of our field study do not meet the ideal requirements of statistics (see below). Our results, therefore, should not depend only on

---

1. Altmann (1974) refers to this sort of sampling as One-Zero scores. I am aware that One-Zero scores measure neither the "true" frequency nor the duration of behaviors, but this is not what I want to do. My aim is to make statements about the "quantum" (the "how much") of behavior; and there is no a priori evidence that frequency or duration tells more about the "how much" of behavior than do the One-Zero scores.

Table 9. Terms of Behavioral Categories, Excluding Locomotion and Vocalization

Category	Elements	Definition
Aggressive behavior	Chasing	A runs after B, who flees.
	Fighting	A aims bites at the shoulder of B, who responds in the same way.
	Lunging at	A runs with raised brows towards B.
	Opening mouth	A directs yawning and staring at B.
	Pumping cheeks	A opens the mouth with the lips forming a circle and covering the teeth, then closes it again. The movement is performed about three times per second.
	Raising brows	A looks at B and lifts the brow skin, displaying the white part of the upper lids.
	Slapping ground	A beats the ground with one or both hands.
	Staring	A protrudes his muzzle and looks at B.
Grooming		A divides and moves the hair of B with one or mostly with both hands.
Infant behavior	Clinging	A holds onto the sides of B.
	Sucking	A has nipple of B in his mouth.
Maternal behavior	Carrying	A transports B on the belly or on the back.
	Embracing	A puts his arms around body of B.
	Grabbing	A grasps and pulls B toward himself.
	Invitation to carry	A presents his back with lowered hindquarters to B.
Nonsocial play		A grasps, pulls, pushes, bites, or rubs inanimate object (stone, sand, root, wood, leaf).
Notifying		A approaches B slowly, looks at him, suddenly turns, presents his face or anal field, and retreats hastily.
Social play		Sequence of mutually and vigorously performed interactions, including pulling, pushing, biting, wrestling, tumbling. Excluded are all sexual and aggressive behaviors.
Scratching		A moves his fingertips at least twice through his fur.
Self-grooming		A grooms part of his own body.
Sexual behavior	Inspecting or touching anal field	A sniffs at or touches briefly the anogenital region of B, who presents.
	Mounting	A stands bipedally and puts hands on B's flanks.
	Penis grip	A touches erected penis or scrotum of B with hand.
	Presenting	A orients his anal field for at least one second toward B.
Wiping muzzle		A briefly touches his nose with a turning movement of his arm.
Yawning		A opens his mouth without aiming the movement at any conspecific.

A indicates the actor, B the recipient of a social interaction.

the application of statistical tests, however significant they may turn out to be. The following means may help to avoid excessive dependence on statistics: (1) The same phenomenon may be measured by independent samples. The agreement of results may be called *convergent consistency*. (2) All results can be tested as to their compatibility with a single model. If this succeeds, the results have *connective consistency* and gain in probability.

In this chapter frequencies of behavior will be investigated for convergent consistency among the broad sample and the close study data. The quasi-independent one-minute records of the broad sample do not suffer from possible effects of cross-interval correlations or from correlations induced by individuals that changed from one age class to the next one during the study period, as in the close study. On the other hand, the close study provides estimates of interindividual variation that cannot be derived from the broad sample.

Broad sample and close study have been analyzed under the following common aspects:

The age factor: What is the influence of age on the frequency of each behavioral category—that is, on the number of intervals with at least one occurrence of the behavior? The answer requires the analysis of the frequency distribution across age classes for each sex. In the case of social behavior, the age distributions were analyzed separately for interactions with partners of their own and of the other sex. In the broad sample the frequencies of each category were tested by the chi-square method against the null hypothesis of no age-correlated difference. If the null hypothesis was rejected, the frequency peak was defined by qualitative inspection of the frequency distribution.

In the close study the test took the form of an analysis of variance in a one-way layout. A significant F-value rejected the null hypothesis. The frequency peak was then defined as significant linear contrast according to the method of Scheffe (in Pfanzagl 1968). For the analysis of variance the frequencies of behavior, Fb, were transformed to $\sqrt{Fb}$ or to $\sqrt{1 + Fb}$, in order to approximate equal variances within the age classes.

The sex factor: What is the influence of sex on the frequencies of each behavioral category? We do not have sufficient data to answer the question for single age classes; therefore we have investigated sex differences that remained the same throughout age classes N to 3—that is, through the age classes that are common to both sexes. The analysis compares the age distributions of males and females for each catagory and tests sex differences against the null hypothesis and against the statistical interactions of sex and age factors. The statistical parameters for the broad sample and the close study are presented in table 10.

Table 10. Survey on the Analysis of Sex-Specific Differences

	Broad sample	Close study
Test design	Contingency table with 2 sexes and 4 age classes	Analysis of variance in a two-layout
Measure of frequency	Number of 1-minute samples with behavior	Percent of 15-second inter with behavior
Frequency of behavior in a sex-age class	Number of 1-minute samples in the sex-age class (no individuals as replicates)	Mean of individual frequencies sex-age class (individuals replicates)
Sex-specific frequency	Sum of age class frequencies per sex class	Mean of age class frequencies sex class
Test of sex difference for independence from age factor	X^2-test for homogeneity in contingency tables	F-test of sex effect aga: interactions of sex and age fa (F_i-value)
Test of sex difference against null hypothesis	X^2-test	F-test of sex effect aga estimation of error (F_e-value)

In the case of social behavior, sex differences result not only from differences between male and female actors, but also from differences of interactions with male and female partners. Figure 9 presents the four possible sex combinations of actors and partners. We tested the following series of comparisons: the frequency of male and female actors (1 + 2 versus 3 + 4); the interactions with partners of the same sex (1 versus 4) and with partners of the other sex (2 versus 3); the interactions with male (1 versus 3) and with female partners (2 versus 4); the interactions of male actors with partners of both sexes (1 versus 2); and the interactions of female actors with partners of both sexes (3 versus 4). These comparisons are interdependent, and three of them are sufficient to describe the

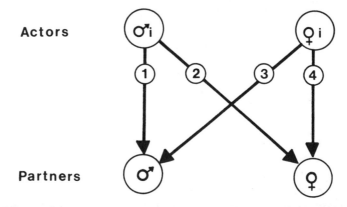

Figure 9. The combinations of actor-partner interactions which are analyzed for sex-specific differences. Actors are immature males and females ($\male_i$, $\female_i$); partners are males and females of all age classes. The actor-partner combinations are numbered; their pairwise comparisons are listed in the text.

differences between the four sex combinations. We tested all comparisons, but in order to save space, I shall present only the interactions of male and female actors with partners of both sexes and the interactions of both sexes with male and with female partners. If nothing else is mentioned, the other comparisons agree with the logical conclusions of the presented results.

To make reading easier in this and the following chapters, all tables with sex differences will contain the same information: (a) the behavioral frequencies of both sexes, (b) the test value for independence of age, and (c) the test value for the sex difference.

The troop factor: The broad sample data of Cone and White Rock permit analyzing differences of frequencies between the two troops. The procedure was the same as for the analysis of the sex factor.

C. Results on Age-dependent Frequencies

The age distributions of frequencies are presented in figure 10 for the broad sample and in figure 11 for the close study. In the broad sample the data of Cone Rock and White Rock were pooled, after we ascertained by qualitative inspection that the distributions were about the same in both troops.

Most naturally, we find the maxima of *infant behavior* and of *being mothered* in infants N. Both behaviors do not persist beyond age class 1. Both samples agree in that *social play* of males with female partners appears mainly during the first year of life; then it decreases rapidly in frequency and play becomes increasingly an isosexual activity. In females, this characteristic is much less conspicuous, since play does not persist in high frequency beyond age class 1. In the close study, the distribution of play among males shows a decrease of frequency in age class B. The decrease seems to be due to the fact that our sample of males B contained two individuals whose social environment promoted play with female partners. Convergent consistency is found for *nonsocial play*, which is most frequent in infants N, then decreases steadily and disappears at the end of the juvenile age. As to *sexual behavior*, the sex of partners has not been taken into account in the broad sample because the frequencies are too low. In both samples we find the highest frequencies of sexual behaviour in juvenile males. *Grooming* shows some differences between broad sample and close study. In the broad sample males seem to groom male partners less frequently in the late juvenile and early subadult age, but this decrease might be a random effect due to the small sample size, and it is not confirmed in the close study. Another difference concerns grooming of female partners by males. In the broad sample the maximum is found in adult males; in the close study the highest

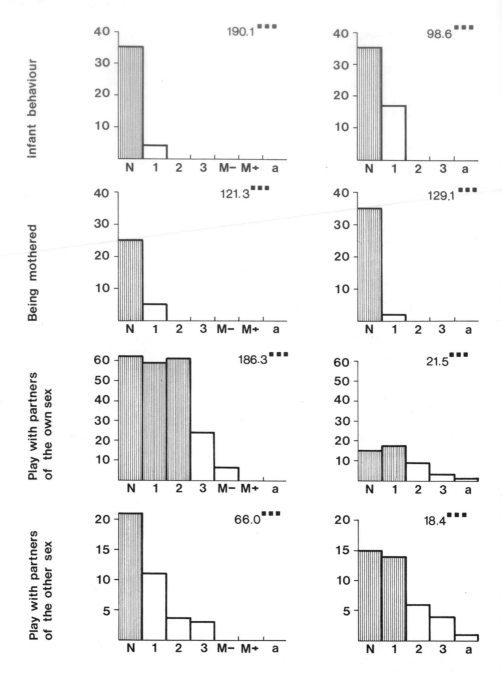

Figures 10a, b, and c. Broad sample: age-specific distributions of behavioral frequencies. The frequencies are based on a total observation time of 12 hours for each sex. The data of Cone and White Rock have been pooled. Frequencies are given as numbers of one-minute samples with at least one occurrence of behavior. The maximum possible score is 120 for each behavior and each age class. The figures in the graphs are X^2 which test the age distribution against the null hypothesis. * indicates a $p < 0.05$, ** a $p < 0.01$ and *** a $p < 0.001$. If the null hypothesis is rejected ($p < 0.05$), the age-specific maximum of frequency is determined by qualitative inspection and represented by hatched bars.

Males

Females

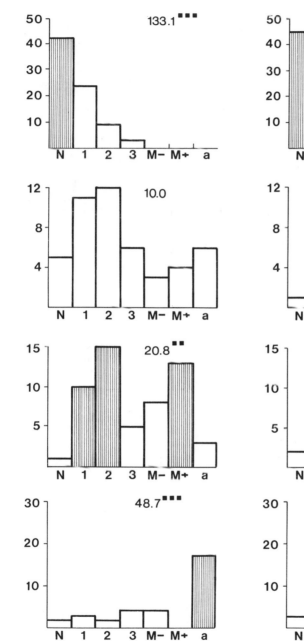
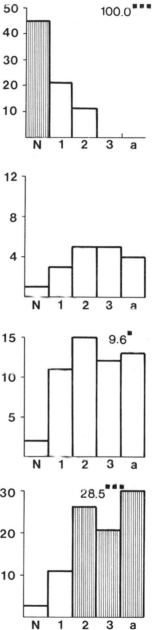

Males   **Females**

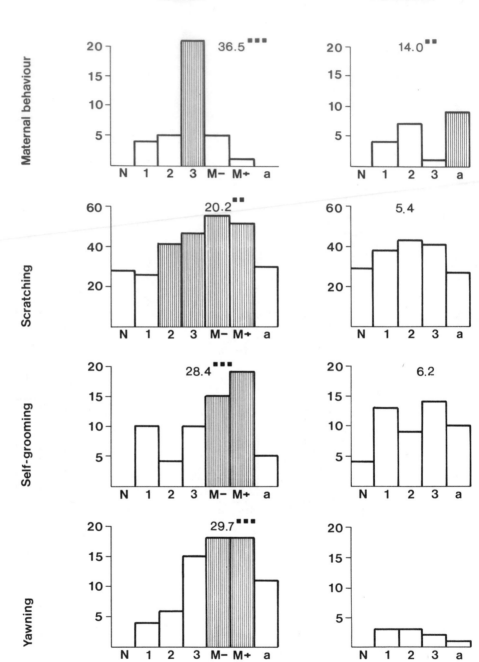

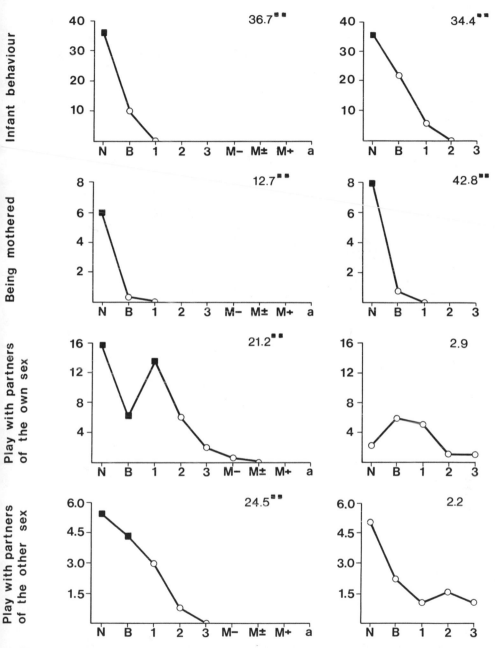

Figures 11a, b, and c. Close study: age-specific distributions of behavioral frequencies. The data are based on a total observation time of 137 hours for males and 59.5 hours for females. Frequencies are given as percents of 15-second intervals with at least one occurrence of behavior. Each point in the graphs represents the mean frequency of several individuals (for the number of individuals per sex-age class see table 2). The figures in the graphs are F-values which test the age distribution against the null hypothesis. ■ indicates a $p < 0.05$, ■ ■ a $p < 0.01$. If the null hypothesis is rejected ($p < 0.05$), the age-specific maximum of frequency is determined as significant linear contrast according to Scheffe and represented by black squares.

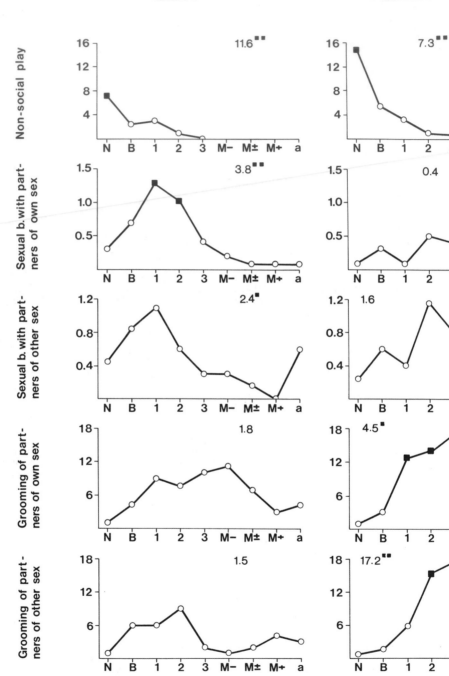

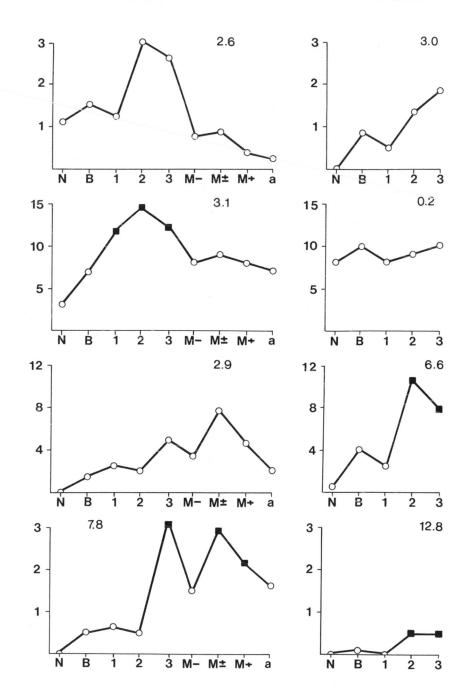

Males

Females

Maternal behaviour

Scratching

Self-grooming

Yawning

frequencies are in juvenile males. The reason for the difference in adult males is that in the broad sample the adult males include the harem leaders, but in the close study only young adult males without one-male units. One possible explanation for the difference in juvenile males is that our close-study sample of males 1 and 2 contained one male, Schnügg, with a particularly high grooming frequency (see Section 4E).

Maternal behavior of males was most frequent in the late juvenile age and of the females in adulthood. *Scratching* in females varied little across age classes. In the close study there was a maximum of frequency in the juvenile age classes of males; in the broad sample the central tendency is found in the late juvenile and subadult age classes. *Self-grooming* and *yawning* were most frequent in subadult males. In females significant maxima of frequency are found only in the close study.

The overall pattern that emerges from the age distributions of frequencies may be summarized in the following way. The development of male behavior proceeds in three phases. The first phase covers the infant and early juvenile age and is characterized by a decreasing attachment to the mother and by frequent play, sexual, and grooming interactions with partners of both sexes. The second phase comprises the late juvenile and the subadult age classes. Its typical feature is a general reduction of social interactions, which consist mainly of grooming with other males. At the same time self-grooming and yawning become more frequent. Finally, the third phase is represented by the adult males and shows increased sexual and grooming interactions with female partners. This pattern of development is in agreement with the results of Kummer (1968a, pp. 84–97), which are based on the study of detached parties and of the nearest neighbors.

In females we may discern a first phase, which comprises the age classes N–1. It is marked by a decreasing attachment to the mother and by few social interactions apart from play with partners of both sexes. Sexual and grooming interactions occur mainly in the second phase, which lasts from age class 2 until adulthood.

D. Results on Sex-dependent Frequencies

The data on sex-dependent frequencies of behavior are given in table 11 for the broad sample and in table 12 for the close study. Figure 12 summarizes the results from both samples.

Infant behavior and being mothered: The close study, as well as the broad sample of Cone Rock and White Rock, showed no age-independent sex difference in infant behavior. But in all samples infant behavior was two to three times more frequent in females B than in males B (see figs. 10

Table 11. Broad Sample: Sex-Specific Differences of Frequencies in Age Classes N to 3

Behavior	Sex of Partner	Rock	Frequencies of actors		Independence of age factor (X^2, d.f. 3)	Sex difference (X^2, d.f. 1)
			Males	Females		
Infant behavior		Cone	18	25	4.3*	1.2
		White	22	27	2.3	0.5
Being mothered		Cone	14	16	1.5	0.1
		White	16	21	0.7	0.7
Play	Own	Cone	84	22	11.6**	——
		White	112	20	3.1	64.1***
	Other	Cone	17	13	2.1	0.5
		White	22	27	3.1	0.5
	Male	Cone	84	13	2.5	52.0***
		White	112	27	1.6	52.0***
	Female	Cone	17	22	7.1	0.6
		White	22	20	10.2*	0.1
Nonsocial play		Cone	42	28	3.7	2.8
		White	40	49	5.9	0.9
Sexual behavior	Own	Cone	13	3	0.2	11.1***
		White	6	0		
	Other	Cone	5	3	6.2	0.2
		White	7	7		
	Male	Cone	13	3	0.5	2.8
		White	6	7		
	Female	Cone	5	3	4.8	5.0*
		White	7	0		
Grooming	Own	Cone	15	24	4.5	1.0
		White	17	16	0.3	0.03
	Other	Cone	7	44	7.2	26.8***
		White	3	18	0.4	10.2**
	Male	Cone	15	44	2.8	14.3***
		White	17	18	4.8	0.03
	Female	Cone	7	24	3.9	9.3**
		White	3	16	1.1	8.4**
Maternal behavior		Cone	13	7	8.2**	——
		White	19	5	3.7	8.2**
Scratching		Cone	69	72	0.5	0.1
		White	75	80	6.0	0.2
Self-grooming		Cone	11	22	1.4	3.7
		White	14	18	2.1	0.5
Yawning		Cone	12	3	3.1	8.8**
		White	13	5		

The terms of frequencies and test values are presented in table 8. A significant X^2 for independence means that the sex effect is not independent of age. * indicates a $p < 0.05$, ** a $p < 0.01$, *** a $p < 0.001$. A graphical summary of this table is presented in figure 2.

Table 12. Close Study: Sex-Specific Differences of Behavioral Frequencies in the Age Classes N to 3

Behavior	Sex of partner	Frequencies of actors		Independence of age factor (F-value) (d.f. 1.4)	Sex difference (F-value) (d.f. 1.3)
		Males	Females		
Infant behavior		9.1	13.1	3.79	3.65
Being mothered		2.5	2.7	2.67	1.67
Play	Own	9.1	2.8	9.42*	24.80**
	Other	2.8	2.2	0.38	0.73
	Male	9.1	2.2	15.41*	32.45**
	Female	2.8	2.8	0.28	0.59
Nonsocial play		2.4	5.4	13.16*	7.35*
Sexual behavior	Own	0.8	0.2	14.23*	14.23**
	Other	0.7	0.7	0.01	0.03
	Male	0.8	0.7	1.06	1.15
	Female	0.7	0.2	5.19	6.81*
Grooming	Own	6.2	9.3	2.67	1.17
	Other	4.5	8.2	1.90	7.44*
	Male	6.2	8.2	1.11	•1.80
	Female	4.5	9.3	3.74	4.78*
Maternal behavior		1.5	0.8	53.33**	6.15*
Scratching		9.9	9.3	0.06	0.40
Self-grooming		2.5	5.4	4.81	9.80**
Yawning		1.0	0.2	2.91	22.86**

The terms of frequencies and test values are presented in table 8. A significant F_i-value means that the sex effect is independent of age. * gives a $p < 0.05$, ** a $p < 0.01$. A graphical summary of this table is presented in figure 12.

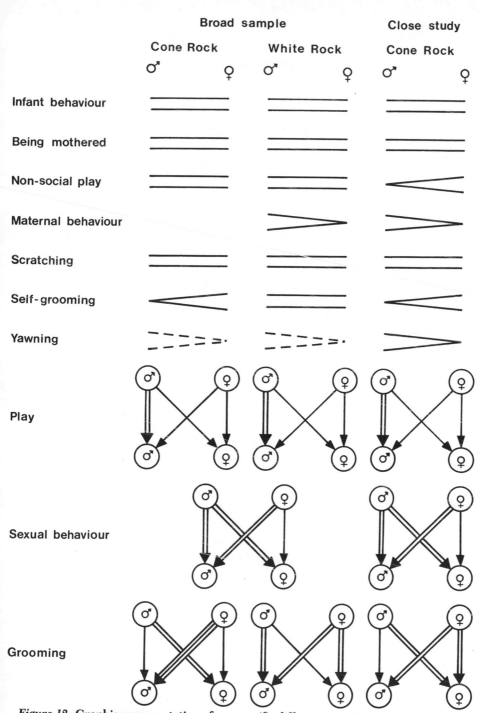

Figure 12. Graphic representation of sex-specific differences in the broad sample and close study. The figure summarizes the results of tables 11 and 12. ♂ = ♀: no sex difference (p > 0.05); ♂ > ♀: behavior more frequent in males (p < 0.05); ♂ < ♀: behavior more frequent in females. The broad sample frequencies of yawning have been pooled for Cone and White Rock and marked with dotted lines. The graphs of the lower part correspond to figure 9. Within each sample and each behavior a triple arrow indicates a frequency which is significantly higher than a double arrow, which in turn is higher than a simple arrow (p < 0.05).

and 11). This difference accounts for the significant heterogeneity in the broad sample of Cone Rock (tab. 11).

In all samples, being mothered occurred with almost equal frequency in both sexes. Thus, while females B seem to perform more infant behaviors, they do not receive more maternal care than males B. This agrees with Kummer's (1968a, p. 91) findings on black infants. We assume that the greater frequency of infant behavior in females B is produced by the initiative of the females B and not by the mother.

Social play: All data demonstrate clearly that males played more frequently than females. Moreover, males preferred male to female partners, while females had no preference as to the partner's sex. Males and females played with partners of the other sex in equal frequencies. The latter result must be expected, because, as we have defined it, play is a mutual interaction. Play with partners of the same sex was not homogeneously distributed across age in the broad sample of Cone Rock (tab. 11) because the sex difference was unproportionally high in age class 2 (fig. 10).

In the broad sample we have divided play into two forms: *play-fighting* involves two animals that face each other with open mouths and fence with their heads by aiming biting movements at each other. Play-fighting, therefore, is the morphological equivalent of actual fighting. All other forms of play that did not conform to play-fighting were referred to as *wrestling*. This distinction was made in the one-minute records of the broad sample, but not in the close study, because the inter-observer reliability decreased after a few minutes of continuous recording. The sex-dependent frequencies of play-fighting and wrestling are summarized in table 13. The figures demonstrate that wrestling occurred with similar frequencies in both sexes; but play-fighting was predominantly a male activity, and it contributed substantially to the higher play frequency in males.

Nonsocial play: A sex-specific difference of frequency was found only in the close study. The higher frequency of females was particularly marked in age classes N and B (fig. 11). This agrees with the qualitative observation of Kummer (1968a, p. 95) that females that avoid dense play groups tend to explore the ground.

Sexual behavior: The frequencies of sexual behavior were so low that the figures of Cone and White Rock were pooled. Sexual behavior occurred with about equal frequencies in all sex combinations, except in female-female interactions, which showed a lower frequency. As to sexual behavior with female partners the sex factor proved significant against the estimation of error, but not against the interactions of sex and age factor

Table 13. Broad Sample: Sex-Specific Differences in Play Fighting and Wrestling

Behavior	Rock	Frequencies of actors		Independence of age-factor	Sex difference
		Males	Females	(X^2, d.f. 3)	(X^2, d.f. 1)
Wrestling	Cone	57	44	4.4	1.7
	White	44	33	5.4	1.6
Play fighting	Cone	54	4	0.8	43.1***
	White	90	14	0.1	55.1***

For the definitions of wrestling see text. *** means a $p < 0.001$.

in the close study (tab. 12). The reason is that the sex difference occurred only in the age classes N-2, but not in age class 3. The residuals of age class 3 contributed nearly two thirds to the total amount of the statistical interactions.

Grooming: There is convergent consistency in the findings that females groomed more frequently than males and that females more frequently groomed partners of the other sex and female partners than did males. Grooming with partners of the same sex showed no significant sex difference, but the frequencies tended to be higher in females at Cone Rock. Males tended more frequently to groom male than female partners, but the difference was significant at White Rock only.

Maternal behavior: Maternal behavior was more frequent in males. In the broad sample of Cone Rock the frequencies were not distributed homogeneously across the age classes (tab. 11) due to the marked peak of frequency in males 3 (fig. 10).

Scratching, self-grooming, yawning: Scratching occurred without sex difference. Self-grooming was more frequent in females of the close study. Because of the small figures for yawning, the broad samples of Cone and White Rock have been pooled. Yawning was more frequent in males.

Figure 12 summarizes the results on sex-dependent frequencies in immature animals. The analysis of the sex factor permits drawing the following conclusions: (1) there are sex differences of behavior that remain consistent throughout age classes N to 3; (2) many of the differences appear convergently in more or less independent samples; (3) behaviors like play-fighting and yawning, both of which are qualitatively associated with aggression, are more frequent in males. Behaviors that are directed to inanimate objects or to their own body, such as nonsocial play and self-grooming, tend to be performed more frequently

Table 14.   Broad Sample: Troop-Specific Differences of Frequencies in the Behavior o
            Immatures

Behavior	Sex of partner	Sex of actor	Troop specific frequencies		Independence of age (X^2)		Troop-specifity $(X^2,$ d.f. 1)
			Cone	White			
Play	Own	Males	84	119	11.2*	(d.f. 4)	6.0*
		Females	22	21	14.3***	(d.f. 3)	—
	Other	Males	17	23	4.7	(d.f. 3)	0.9
		Females	13	27	3.4	(d.f. 3)	4.9*
Nonsocial Play		Males	42	40	4.0	(d.f. 3)	0.1
		Females	28	49	4.8	(d.f. 3)	5.7*
Grooming		Males	20	11	1.0	(d.f. 3)	2.6
		Females	62	30	3.3	(d.f. 3)	11.1***

by females. The same is true for grooming, which one would consider a
friendly behavior; (4) there is no evidence that these differences are
correlated with sex-specific mothering of infants, since being mothered
occurred with about equal frequency in both sexes. Yet there remains
the possibility that our measures of being mothered were not sufficiently
differentiated.

E. Results on Troop-dependent Frequencies

The broad sample data of Cone Rock and White Rock were generally
in agreement. Yet there were some significant differences of frequency
between the troops. They are summarized in table 14. The behaviors
not included in the table show no significant troop difference.

The important difference is that Cone Rock females up to age class 3
groomed male partners more frequently than did White Rock females.
This difference refers particularly to mature male partners: Cone Rock
females were found to groom mature males in 36 samples, White Rock
females only in 12 samples. We assume that grooming interactions between immature females and mature males represent what Kummer
(1968a) has described as *initial units*. It seems from our data that either
the grooming interactions within initial units were less frequent at White
Rock or that the probability of juvenile females being members of an
initial unit was higher at Cone Rock. Qualitative evidence suggests that
initial units were actually less frequent at White Rock. If this is correct,
we may ask the question: What do females 2 and 3 do while they are not
socially bound to an initial unit? Table 14 shows that White Rock females played more frequently with male partners than did Cone Rock
females; they also were more often engaged in nonsocial play. The lack
of homogeneity in the age distributions of play with male partners (tab.
14) is produced by a marked troop difference in age class 2: Cone Rock
females 2 rarely played with male partners; instead they frequently

groomed mature males, compared to White Rock females of the same age.

Furthermore, young subadult males of White Rock were found in 7 one-minute samples to play with male partners, whereas no such interaction was recorded at Cone Rock. This difference accounts for the heterogeneity in the age distributions (tab. 14). Young subadult males are not yet initial leaders, but they represent the age that immediately precedes the stage of initial units. If in the young subadult age there are any social processes that finally lead to the establishment of an initial unit, these processes might suppress play interactions with male partners.

We hypothesize: life in the initial unit suppresses social and nonsocial play in juvenile females; play in males is inhibited in the age that precedes the establishment of the initial unit. This integrates all observed troop differences into a consistent hypothesis, which will be discussed again in chapter 5.

Ontogeny of Social Relationships

The last chapter has tested behavioral frequencies against the null hypothesis. Rejection of the null hypothesis, however, tells us nothing but that the results are not random. What then accounts for the differences, if not mere accident? As a preliminary answer, this chapter attempts to test the compatibility of behavioral frequencies with a systemic model. The model is derived from a pilot study by Kummer (1975) that was carried out on captive geladas, but since then has also been extended to captive hamadryas (Hinderling 1975).

A. The Bonding-Stage Model

The basic result of Kummer's pilot study is the finding that social bonds among adult geladas and hamadryas develop in a regular sequence of first occurrences: in an encounter of two animals who meet for the first time or after a long separation, the first fight, if it occurs, precedes the first presentation, which in turn precedes the first mounting; the first mounting, finally, precedes the first grooming. This is the rule of the *stage sequence* in bond development.

In dyadic encounters the speed of the stage sequence is correlated with the status combination of the animals involved: the more equal in status the animals are and the higher the status, the more probable the occurrence of fighting and the more time needed to reach the stage of grooming, if it is reached at all. On the other hand, the greater the differences of status and the lower the status, the lower the probability of fighting and the sooner the animals groom. The speed of the stage sequence is preliminarily called the compatibility of the dyad. Compatibility is correlated negatively with the behavioral frequency of the first bonding stage and positively with the behavioral frequency of the last bonding stage, but not with the frequencies of the middle stages (Kummer personal communication; Hinderling 1975). These are the *dyadic rules* of bond development.

The purely dyadic development of social bonds is normally possible only in the experimental situation. In a social group each pair bond is affected by the presence and behavior of the other group members. The

70

simplest case of such a situation is the triad. The characteristic feature of the triad is that at least one of the three pair bonds does not reach the same stage or the same behavioral frequency of a stage as in the dyadic situation. This means, in the terminology of Kummer, that in the triad at least one pair bond regresses. The regression may be effected by the following means: (1) By *intervention*. Social intervention in hamadryas can assume one of three forms: a. *interposition:* a third animal can place himself between two potential partners of a dyad and thus prevent their interaction; b. *exclusion:* a third animal may actively exclude one of the dyad's partners by inducing its withdrawal through threat or attack; c. *herding:* a third animal may actively herd a dyad member away from its partner by inducing its approach or following reaction. (2) By *inhibition*. Within a triad a dyadic interaction can have an inhibitive effect on the social behavior of the third animal without resort to intervention. The rules that describe the relationships within a triad will be referred to as *triadic rules* of bond development. Their systematic investigation has not yet been completed in hamadryas. The only tested rule derives from the experiments of Kummer et al. (1974) and says that a male-female bond inhibits the behavior of a male rival toward the pair.

B. Methodical Approach

The bonding-stage model was tested on the close study data of immature males and females. The partners of the immatures were classified as immature males, immature females, and mature males. These classes of partners occurred roughly in equal proportions within the troop (see tab. 3). Two classes of actors and three classes of partners provide us with six sex-age combinations of dyadic interactions. For testing the dyadic rules we assumed a status order of mature males > immature males > immature females. The assumption that immature males are higher in status than immature females is supported by the following close-study data on heterosexual encounters of immatures: females had an approach/withdrawal ratio of 0.9 toward immature males, males a ratio of 1.3 toward immature females (sample size = 336, X^2 = 6.68, p < 0.01); the ratio of jumping at partner/jumping away from partner was 0.6 for females and 2.5 for males (sample size = 65, X^2 = 12.27, p < 0.001); of 44 instances of chasing-fleeing sequences, males were the chasers 41 times, and in 61 instances of aggressive interactions (staring at, lunging at, or biting partner), males were the aggressors 51 times.

Our field data required some modifications of Kummer's approach:

a. Kummer's subjects were isolated dyads, which were formed by uniting strangers for the first time. The development of the dyads was followed up continuously. Our data describe interactions in already existing relationships within a free-ranging society. The ontogenetic development of these interactions is reconstructed by a cross-sectional approach. The lack of continuous records on single dyads made it impossible to use temporal measures such as the first occurrence of a behavior or the speed of a sequence. Instead, we had to apply frequency measures of bonding stages, which are defined in the following way: (1) a particular sex-age combination is said to have attained a bonding stage when the stage-specific behavior has reached its ontogenetic maximum of frequency; (2) a particular sex-age combination is said to be more compatible than another one if the behavior of the lowest stage is less and that of the highest stage more frequent in the dyads.

b. Kummer studied subadult and adult animals, while we focused on immature animals. In immatures we found no fights, which represent the first bonding stage in Kummer's concept. The morphological equivalent of fighting is what I have defined as play-fighting, which, however, was recorded only in the broad sample as a distinct element (see chapter 3). For the close study we had to take the general definition of play as a substitute for fighting. This step seems acceptable because differences of play frequencies were shown to result mainly from differences of play-fighting frequencies (see tab. 13). Accordingly, play is defined here as the first bonding stage.

Our frequencies for presenting and mounting, which represent the second and third bonding stages in Kummer's study, were too low for separate investigations. Therefore we had to pool presenting and mounting. The second bonding stage is referred to as sexual behavior. Since sexual behavior is not used as a measure of compatibility (cf. above), such pooling of elements is not a far-reaching modification.

Grooming, Kummer's fourth bonding stage, appears here as the third stage.

With these definitions we ask the following questions:

1. Do the ontogenetic maxima of play, sexual behavior, and grooming in interactions of immatures follow the same sequence as the first occurrences of bonding stages in isolated dyads of adult strangers? The answer to this question decides whether the further predictions of the bonding-stage model may be tested on the behavior of immatures.

2. Do differences of play and grooming frequencies in various sex-age combinations agree with the predictions of the dyadic rules? An affirmative answer permits interpretation of some differences of behavioral frequencies as the result of differing dyadic compatibilities in immatures.

3. If we find deviations from the dyadic rules, is it possible to relate them to the influence of third parties?

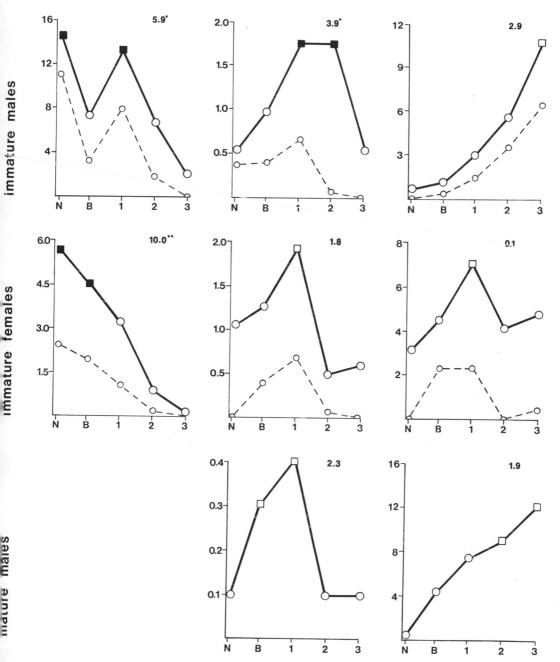

Figure 13. Frequencies of play, sexual behavior and grooming, received and given, in social interactions of immature males N to 3. The data are based on a total observation time of 55.5 hours. The age-specific distributions are presented across the age classes of the reference animals for interactions with immature males, immature females, and mature males (———), and for interactions with immatures of the subject's own age class (– – –). Each point represents the mean frequency of 4 individuals as percents as in figure 11. The figures in the graphs are F-values which test the age distribution against the null hypothesis; ■ indicates a $p < 0.05$, ■■ a $p < 0.01$. Black squares represent significant, empty squares insignificant age-specific maxima of frequencies, which are defined as linear contrast according to Scheffe.

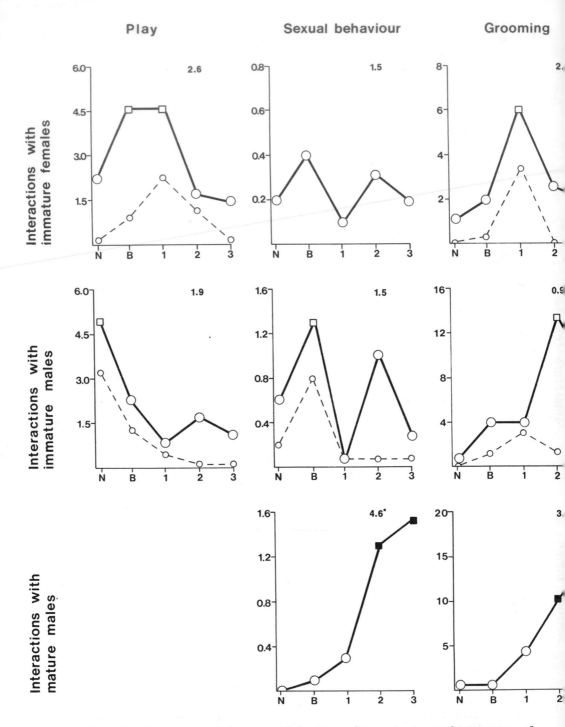

Figure 14. Frequencies of play, sexual behavior, and grooming in social interactions of immature females N to 3. The data are based on a total observation time of 59.5 hours. Each point represents the mean frequency of 3 to 5 individuals as percents (see tab. 2). For further explanation see figure 13.

C. The Maxima Sequence of Play, Sexual Behavior, and Grooming

In the first step we investigate the maxima sequence of bonding stages during ontogeny on the close-study data. The frequency distributions of play, sexual behavior, and grooming are presented in figure 13 for dyadic interactions of immature males and in figure 14 for interactions of immature females. Play of immatures with mature males is not given, because the frequencies are too low.

For each age class of actors, figures 13 and 14 give the frequency of interactions with the three partner classes: immature males, immature females, and mature males. For interactions among immatures, it would be best to differentiate the partner classes further and to analyze interactions among animals of the same age class—that is, among animals who have been growing up together. But our respective frequencies are too low and interindividual differences too great for a quantitative analysis. Nevertheless, I have added the frequencies for interactions among animals of the same age (dotted lines in figures 13 and 14) in order to show qualitatively that our broader partner classes provide essentially the same frequency distributions.

The maxima sequence of play, sexual behavior, and grooming is in agreement with the bonding-stage model if it meets the following condition: the maximum of play occurs in an earlier or in the same age class as the maximum of sexual behavior, which occurs in an earlier or the same age class as the maximum of grooming. The occurrence of maxima in the same age class does not contradict the bonding-stage model because we cannot expect that the transition from one bonding stage to the next one always coincides with a change of age class. Figures 13 and 14 show that the condition is met in all sex-age combinations of dyadic interactions.

How probable is the chance occurrence of a maxima sequence that agrees with the bonding stage model? As I define it, the bonding stage is reached with the *first* age class that belongs to the maximum of frequency. With three bonding stages, each of which may be reached in one of five age classes, 125 random combinations of class-maxima are possible. Thirty-five of these combinations meet the above conditions for the agreement with the bonding-stage model. Accordingly, the probability for a random sequence to agree with the bonding-stage model is $p = 0.28$ in each sex-age combination of dyadic interaction. If the sequences of maxima in various sex-age combinations are completely independent of each other, the probability of six sequences to agree with the bonding-stage model is $p = 0.2^6 = 0.0005$. If the sequences of various sex-age combinations were not independent, an estimate of $p = 0.2$ for the first and $p = 0.6$ for the following sex-age combinations would lead to a total probability of $p = 0.02$ for six sex-age combinations to be in agreement with the model. These estimates support my conclu-

Table 15. Differences of Compatibility between Various Sex-Age Combinations

Comparisons of sex-age combinations		Play			Grooming		
		Frequency	F_i	F_e	Frequency	F_i	F_e
1	Imm.M with imm.M	8.7	9.19*	25.29**	4.3	0.26	0.59
	Imm.F with imm.F	2.8			2.7		
2	Imm.M with imm.M	8.7	19.77*	31.48**	4.3	0.27	0.34
	Imm.M with imm.F	2.8			4.8		
3	Imm.F with imm.F	2.8	1.08	2.15	2.7	2.00	1.47
	Imm.F with imm.M	2.2			5.7		
4	Imm.M with imm.F	2.8	0.33	0.68	4.8	0.12	0.06
	Imm.F with imm.M	2.2			5.7		
5	Imm.F with imm.M	2.2	24.09**	33.61**	5.7	0.07	0.06
	Imm.F with mat.M	0.04			6.5		
6a	Imm.M with mat.M	0.4	4.25	2.43	6.7	0.88	0.63
	Imm.F with mat.M	0.04			6.5		
6b	Mat.M with imm.M	0.1	9.17	5.79*	2.0	0.11	0.23
	Mat.M with imm.F	0			2.6		
7	Imm.F with imm.F	2.8	20.36**	51.69**	2.7	0.56	1.44
	Imm.F with mat.M	0.04			6.5		
8	Imm.M with imm.M	8.7	36.19**	155.00**	4.3	3.71	11.14**
	Imm.F with mat.M	0.04			6.5		
9	Imm.M with imm.M	8.7	29.09**	104.73**	4.3	7.86*	1.88
	Imm.M with mat.M	0.4			6.7		

Each comparison lists first the sex-age combination which is predicted to be less compatible than the second combination (except comparison 4, see text for further comments). Frequencies are means across age classes, calculated from figures 13 and 14. Imm. = immature; mat. = mature; M = males; F = females; F_i is the F-value, which tests the difference of frequency against the statistical interactions of sex-age combination and actor class; F_e tests the difference against the error; * represents a $p < 0.05$; ** a $p < 0.01$.

sion that the agreement of the ontogenetic sequences of maxima with the bonding-stage model is unlikely to be based on mere chance. Thus the most fundamental feature of the model occurs also in the ontogenetic development of social relationships in immatures. This encourages us to proceed and to investigate the validity of the dyadic rules.

D. The Dyadic Rules

The dyadic rules of bond forming state that a dyad is the more compatible the greater the difference of status and the lower the status of the animals involved. In our data a high compatibility is characterized by a low play and a high grooming frequency. The dyadic rules were derived from observations of short-term relationships under experimental conditions. In order to apply them meaningfully to long-term relationships within a free-ranging society I make the following assumption: a dyad regresses from a bonding stage or a stage frequency it has once attained only under the influence of triadic effects, but not under the influence of increasing age, which leads to a higher status. Two males, for example, who have reached the grooming stage as juveniles will continue to groom each other as adults, unless they are affected by triadic effects; in contrast, the original dyadic rules would predict among mutually unfamiliar adult males that grooming is unlikely to occur. The above assumption, therefore, provides a tentative assessment of the long-term effect of familiarity and a criterion for recognizing possible triadic effects.

The dyadic rules are tested by pairwise comparison of play and grooming frequencies in all sex-age combinations whose difference of compatibility is predictable from the bonding-stage model. The comparisons are presented in table 15. Whenever the observed frequencies do not agree with the predictions, I shall refer to the age-specific frequency distributions in figures 13 and 14 and deduce a hypothetical triadic rule that could explain the deviation from the dyadic rules. The postulated triadic rules will represent the starting point for the following chapters on family forming. The main results on the dyadic rules are graphically summarized in figure 17 at the end of the chapter.

The comparisons in table 15 are presented so that the sex-age combination that is listed first in the comparison is predicted to be less compatible than the second combination. An exception is comparison 4 (see below). The sex-age combination listed first should be less compatible because either the absolute statuses of its members are higher (comparisons 1, 2, 6) and/or the status difference is smaller (comparisons 2, 6, 8). At present the bonding-stage model cannot yet predict the difference of compatibility if one dyad contains animals with low status and the other dyad contains animals with a great difference in status. In adult geladas,

Table 16. Age Combination in Isosexual Play Encounters of Immature
           Animals

Sex of actor	Age classes of actors	Age combination of play interaction	Frequency	F_i	F_e
Males	N to 2	Play with partners of same age	5.8	7.63	23.04**
	N to 2	Play with older partners	3.0		
	B to 3	Play with partners of same age	4.1	0.65	10.53**
	B to 3	Play with younger partners	2.1		
Females	N to 2	Play with partners of same age	1.4	0.01	0.45
	N to 2	Play with older partners	1.0		
	B to 3	Play with partners of same age	1.3	0.09	0.20
	B to 3	Play with younger partners	1.1		

The table compares isosexual play interactions between partners of the
same and of other age classes. F_i tests the difference of frequency
against the statistical interactions of age combination and actor age,
F_e against the estimation of error.   ** indicates a $p < 0.01$.   The
table shows that males prefer to play with male partners of the same
age class.

Table 17. Age Combinations in Heterosexual
           Play Encounters of Immature Animals

Play with	Frequencies of actors		F_i	F_e
	Males	Females		
Older partners	1.3	0.4	3.87	15.63**
Partners of same age	1.0	1.1	0.20	0.03
Younger partners	0.2	0.7	0.90	4.50*

The table compares age combinations in heterosexual play encounters between male and female
actors N to 3. F_i tests the difference of
frequency against the statistical interaction
of actor sex and actor age, F_e against the
estimation of error; * indicates a $p < 0.05$,
** a $p < 0.01$.   The table shows that play
between younger males and older females is
more likely than vice versa.

however, male-female dyads seem to be more compatible than female-female dyads (Kummer 1975). This is an indication that the effect of status difference between the sexes overrides the effect of the low female status. The qualitative impression that the same applies also to our sex-age combinations has been used to predict the outcome in comparisons 3, 5, 7. The prediction for comparison 9, finally, is completely based on qualitative evidence.

1. Interactions among Immatures

Comparisons 1 to 3 in table 15 test the dyadic rules for interactions among immatures. Comparison 4 is a control, because it presents the same sex-age combination, whose frequencies of interaction are derived from the independent male and female records. Here, accordingly, no difference of frequency should occur. In comparisons 1 to 3 all differences of play frequencies agree with the predictions. Only in comparison 3 does the difference appear as a clear tendency without being statistically significant against the error; neither is it significant against the statistical interaction of the sex-age combination and age; here the reason lies in the particularly great difference of frequency in age class 1 (see fig. 14).

The agreement of the play frequencies with the dyadic rules permits us to add a further detail: the dyadic rules predict that play frequencies increase with decreasing status differences. In isosexual play encounters we should therefore expect play to occur mainly among animals of the same age. The expectation is tested in table 16, which compares play with partners of the same age and play with older partners for age classes N to 2 (age class 3 is omitted because here play with older immatures is impossible); then play with partners of the same age is compared to play with younger partners for age classes B to 3. Table 16 shows clearly that in males play is most frequent among animals of the same age. In females the same result is less conspicuous for the following reason: females N played frequently with females B. This increased frequency of play with older partners led to a high figure for the statistical interaction between actor age and partner age (low F_i-value in table 16!). The same effect is produced again by the females B in their play with younger partners.

In heterosexual play encounters, play will also be frequent between animals of the same age—that is, between animals who have been growing up together. But since males are higher in status than females, we should expect play between a younger male and an older female to occur more frequently than between an older male and a younger female. This expectation is confirmed in table 17.

Grooming among immatures is less consistent with the dyadic rules than is play. In comparison 1, females do not groom more among themselves than do males, as the dyadic rules would predict. The frequency distribution of grooming among females (fig. 14) shows that females seem to reach a maximum in age class 1; but then grooming interactions among females do not continue on the same frequency level; instead they decrease during the age classes 2 and 3. In comparison 2, males groom in equal frequencies with male and female partners, whereas the dyadic rules would predict a preference for female partners. In the frequency distributions of figure 13, we find that grooming between males and females appears to decrease during age classes 2 and 3. In comparison 3, the difference of frequencies agrees with the dyadic rules, but it is statistically not significant because of great interindividual differences of frequency. Comparison 4, which serves as a control, shows no difference of frequency, thus agreeing with our expectation.

I interpret the comparisons 1 to 4 as follows: the deviations of the grooming frequencies from the dyadic rules seem to have their origin in immature females, who groom less frequently with immatures of both sexes at the end of the juvenile period. In this period, however, females reach the maxima frequencies for sexual behavior and grooming with mature males (fig. 14). This leads to the hypothesis that interactions between mature males and immature females suppress the females' grooming interactions with immatures. This is the first triadic effect that I have to postulate. A similar hypothesis, however, has already been advanced in the discussion of troop differences in the broad sample. There I have assumed that interactions between mature males and immature females reflect initial units. This first triadic effect, therefore, seems to operate in the second phase of the female's juvenile period, when the female becomes a member of an initial unit. Chapter 5 will confirm that the initial bond suppresses interactions between initial females and immatures (cf. tab. 24). At the same time, it will reveal that not all juvenile females are members of an initial unit; this explains part of the great interindividual differences of frequencies among immature females, which leads to the statistical nonsignificance of the difference in comparison 3.

2. Interactions between Immatures and Mature Males

Comparisons 5 to 9 in table 15 deal with interactions between immatures and mature males. Comparisons 6a and 6b are controls in that they contain the same dyadic interactions; the figures of comparison 6a are derived from the close-study data on immatures, those of comparison 6b from the data on mature males.

The play frequencies are again consistent with the dyadic rules: male and female immatures play much less with mature males than with immature partners (comparisons 5, 7, 8, 9), and males tend to play more

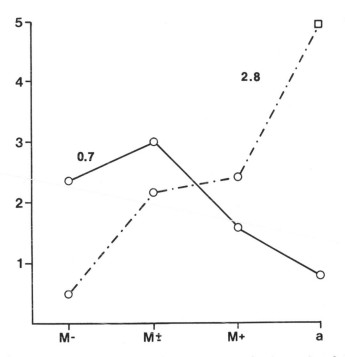

Figure 15. Grooming of mature males with immature males (———) and immature females (—·—·—·). The frequencies are intervals with grooming received and/or given by mature males, expressed as percents of all observation intervals; they are based on a total observation time of 81.5 hours. The number of mature males per age class is found in table 2. The figures in the graph are F-values which test the age distribution against the null hypothesis. The empty square indicates an insignificant maximum of frequency.

frequently with mature males than do females (comparisons 6a and 6b). The consistency with the dyadic rules appears clearly in females, who play in decreasing frequencies with increasing status of their partners (comparisons 3 and 5).

The grooming frequencies deviate from the predictions of the dyadic rules. Comparisons 5 and 6 indicate that the frequency of grooming between immature females and mature males is below the expectation from the dyadic rules. In comparison 7 the difference agrees with the prediction, but it is not statistically significant due to great interindividual differences of frequency. Only in comparison 8, which contains the most and the least compatible sex-age combinations, does a significant difference occur. Comparison 9, finally, suggests that the frequency of grooming between immature males and mature males might be lower than expected, since the difference of frequency is not significant.

What might be the causes that produced these deviations from the dyadic rules? Figures 14 and 15 reveal that grooming between immature females and mature males was most frequent in females 2 and 3 and in young adult males. The dyadic rules, however, lead us to expect the maximum of grooming earlier in ontogeny. I hypothesize, therefore, that interactions between immature females and mature males do not reach the expected frequency of grooming because the ontogenetic development of these interactions is somehow delayed. Kummer (1968a, pp. 92–93) has reported in his results on grouping tendencies that subadult males are preferentially found near the troop's periphery, while one-year-old females remain in the vicinity of adult males. It seems natural to postulate the leader's intolerance toward potential rivals as the main reason for the low grooming frequency between young immature females and subadult males. This is the second triadic effect that has to be introduced. It seems to operate during the first phase of a female's life when she still belongs to her parental unit. I shall deal with the leader's influence on interactions between immature females and subadult males in the following chapter. But three observations from Cone Rock troop may demonstrate that subadult males have indeed to face the leader's tendency to exclude him and to herd immature females:

> A female infant approaches a subadult male who sits near a one-male unit. The unit leader immediately comes near, grabs the infant, and carries it back to the one-male unit.

> A subadult male approaches a female B; he grabs her and carries her away. The female's mother shows an intention to follow, but then remains standing and starts screaming while looking at her leader. The leader stares at the subadult, pumps cheeks, and, when the subadult does not respond, lunges at him and chases him away. The released female B returns to her mother.

> A subadult male approaches a female 1 while pumping cheeks. The leader of the female lunges at the subadult, chases after him, and, after a short fight with him, returns to his harem unit.

Finally, we must think of a possible reason that immature males do not groom mature males significantly more often than other immature males (comparison 9 in tab. 15). From figure 15 it seems that young subadult males groom more frequently with immature males than do old subadult and young adult males. Grooming with immature males, then, seems to decrease as soon as grooming with females becomes more frequent. In other words; grooming between mature males and immature females might suppress grooming between mature and immature males. This is the third triadic effect to be postulated from the investigation of the dyadic rules. It would operate as soon as old subadult and young adult males establish their initial unit–that is, at the same time as the first triadic effect postulated above.

Table 18. Sex-Specific Differences in
Grooming between Immatures and
Adult Females

Behavior of actor	Frequencies of actors		F_i	F_e
	Males	Females		
Grooming adult females	2.8	7.9	4.53	6.83*
Being groomed by adult females	9.1	6.0	0.63	1.45

The table gives the grooming frequencies of males and females N to 3 with adult females. F_i tests the difference of frequency against the statistical interaction of actor sex and actor age, F_e against the estimation of error; * indicates a $p < 0.05$. The table shows that females groom more frequently adult females than do males, but that males and females are groomed by adult females in similar frequencies.

E. Interactions between Immatures and Adult Females

Until now, I have ignored interactions between immatures and adult females. The reason is that the bonding-stage model was derived from the behavior of mature animals and therefore has not yet integrated maternal and infant behavior into its concept. Interactions between immatures and adult females are presented here without reference to the bonding-stage model, in order to complete our view on social relationships of immatures.

Besides infant behavior and being mothered, the frequencies of which have already been presented in figures 10 and 11, grooming was the most frequent behavior to occur between immatures and adult females. Figure 16 summarizes the age distributions of grooming. Grooming and being groomed increased in males until age class 2 and then decreased again; in contrast, immature females steadily increased their grooming frequency during the juvenile age, while being groomed remained more or less on the same frequency level. This agrees with the general development that

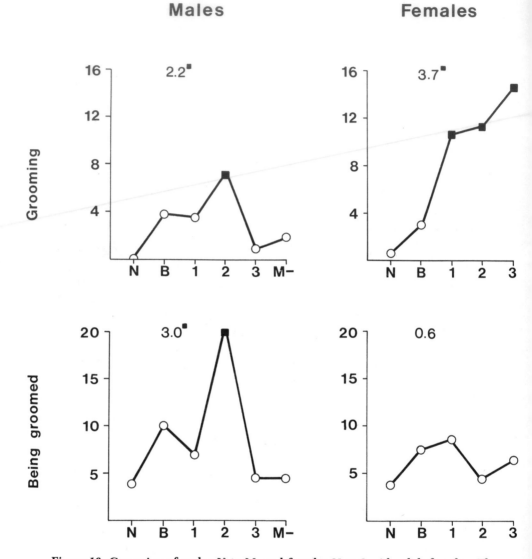

Figure 16. Grooming of males N to M- and females N to 3 with adult females. The frequencies are based on a total observation time of 82.5 hours for males and 59.5 hours for females. The figures in the graph are F-values that test the age distribution against the null hypothesis; ▪ indicates a p < 0.05. Black squares mark significant maxima of frequencies, which are defined as linear contrasts according to Scheffe.

Kummer (1968a) has described: juvenile males tend to leave their mothers' unit as older juveniles and to join other males, whereas females stay in their mother's unit until they are taken over by another leader.

Table 18 demonstrates that immature females groomed adult females more frequently than did males. The difference, however, is not significant against the statistical interactions between sex and age because the sex difference is particularly great in age class 3 (fig. 16). Being groomed occurred without sex difference in immatures. Figure 16 seems to suggest that being groomed is more frequent in males. But the high frequency of being groomed in males 2 was caused mainly by a single male, Schnügg, who was groomed in 43.1 percent of intervals by adult females (see below). Without him, the frequency of being groomed in age class 2 is reduced from 20.5 (fig. 16) to 12.9 percent, and the overall frequency of being groomed in immature males drops from 9.1 (tab. 18) to 7.6 percent of intervals. The sex difference in grooming between immatures and adult females reveals the same pattern that we have already met within infant behavior and being mothered (see above): that immature females interact more frequently with adult females than do males is not an effect of an increased activity of the adult females but is initiated by the immature females. Insofar as the adult females are the mothers of the immatures, this conclusion agrees with the observation of Kummer (1968a, p. 58) that in the mother-child relationship the initiative passes to the child as soon as the black infant coat is lost.

Grooming between immatures and adult females becomes more frequent after infant behavior and being mothered have declined in frequency (figs. 11 and 16). Do these grooming interactions represent a continuation of the mother-infant bond? From our observations we can deduce a few indications toward an affirmative answer:

In immatures we found an overall tendency to groom preferentially with a single adult female. This is confirmed by 5 of 12 marked males N to 2, by 5 of 7 males 3 to M-, and by 10 of 13 females N to 3. Five males N to 2 and 3 females N to 3 groomed each within a single one-male unit whose adult females we could not identify individually. Only one male showed no preference for interaction with a single female. It was Schnügg from band II. Schnügg, a male 1 to 2, groomed regularly with at least three adult females from different one-male units. We had the intuitive impression that Schnügg had lost his mother, but we lack evidence to support the case.

It seems plausible to assume that the adult females with whom immatures groomed were normally the mothers. The assumption, however, will not apply to all cases, because a mother-daughter relationship might be abruptly broken when the mother or the daughter is taken over by a new leader.

The males 3 to M- were noticeable in that the females they groomed with appeared to be very old. In addition, grooming between males 3 to M- and adult females took place in face of an amazing tolerance of the females' leaders. This tolerance contrasts with our hypothesis that harem leaders suppress interactions between subadult males and immature females (see above). In band III a male 3 who did not belong to our close-study sample went so far as to herd an old female and to form a subunit within the harem:

> The male 3 leads the adult female away from her one-male unit, checking her following response by looking back at her. Five meters from the female's leader, they sit down. One minute later, the leader approaches the pair. The female presents to him, but he ignores her. The female then presents to the male 3 who leads her again away from the adult male. The leader does not intervene. Twice the leader passes the adult female and the male 3 with the rest of his harem. Each time the adult female shows an intention to follow, but the male 3 immediately gets up and leads her away from the one-male unit.

Two hypotheses might explain that subadult males groom with old females and that these interactions are tolerated by the harem leaders: either harem leaders are not very possessive of old females, thus making them easily accessible to subadult males, or the subadult males are sons of the old females and the persisting mother-son relationship has an inhibitive effect on the leader's behavior toward the pair. I am inclined to consider the second hypothesis more probable for the following reasons: we never observed an old female to be groomed by more than one subadult male, as would be expected if harem leaders were neglectful of old females. Also Gox, a male 3 in band II, repeatedly groomed with the only adult female of a young harem leader; it seems improbable to me that a harem leader should not be possessive of his only female. Finally, Pepsi, a male M- in band I, regularly groomed with one of the two females that belonged to the young harem leader, Bishop. Bishop tolerated grooming between Pepsi and the old female, but intervened as soon as Pepsi attempted to copulate with the female.

F. Discussion

This chapter was an attempt to apply the bonding-stage model to dyadic interactions within a free-ranging society. The analysis has shown that the sequence of bonding stages also occurs in the social relationships of immature animals. While it takes minutes, hours, and sometimes days to pass through the sequence between two adults newly convened in an experimental situation, it takes months and years in the ontogeny of social relationships. One factor that might contribute to the long duration of the ontogenetic sequence is the fact that the dyadic interactions that we have analyzed occurred in a broad group context. In the experiments of Kummer (1975) a group of 6 or 7 animals increased the time for reaching a given stage by more than a factor of 10.

Possibly the same ontogenetic sequence of play, sexual behavior and grooming occurs also in other primate species. According to the date of Hinde and Spencer-Booth (1967) on rhesus monkeys, the maximum of frequency is reached in the first year for play and in the second year for mounting, while the frequency for grooming was still increasing after 2.5 years.

Further analysis has demonstrated that differences of play frequencies agree with the dyadic rules, which relate the compatibility of dyads to the status combination of the dyadic partners. Play occurred in various sex-age combinations of interactions in accordance with the compatibility of the play partners and did not seem to be notably affected by triadic effects. This does not mean, however, that there are no general preconditions that promote the occurrence of play. Mason (1965a) has demonstrated for chimpanzees that mild novelty promotes play, while increasing unfamiliarity promotes clinging. For an immature animal the familiar surrounding is mainly represented by the mother. In rhesus monkeys the absence of the mother decreases the frequency of play (Kaufmann and Rosenblum 1967, Spencer-Booth and Hinde 1971). The same effect occurs when the mother does not provide adequate and experienced maternal care (Seay et al. 1964, Mitchell et al. 1966). If the results of Mason (1965a) have a general validity, we may hypothesize that females need more security from their mothers in order to play than do males. Possibly females play regularly only as long as they can cling to their mother, whereas grooming with the mother is sufficient for males in order to play. Our observations agree with this hypothesis: females show a stronger tendency toward withdrawal in play groups than do males (see section 31). Female play decreases shortly after infant behavior has disappeared in mother-daughter relationships (see figs. 11 and 14), whereas males still play frequently when their relationship to the mother has changed to the grooming stage (see figs. 13 and 16). The decline of play and of infant behavior in females seems also to coincide in langurs (Sugiyama 1965a) and in squirrel monkeys (Baldwin 1969).

Our substitution of play for fighting as the first bonding stage is a major modification of the bonding-stage model. Play has been used as an estimate for social play-fighting, which is a ritualized form of the actual fight. According to the dyadic rules, fighting is most likely to occur in dyads with low compatibility—that is, in dyads that do not easily reach the stage of grooming. Bond formation in the group context provides an additional difficulty in reaching the grooming stage and increases the frequency of fights. From this aspect it seems efficient to ritualize fights in order to minimize the risk of wounds. In addition, a high frequency of fights that are transformed to play permits animals who have to interact in dyads of low compatibility to be positively reinforced for keeping close distances, as has been suggested by Harlow (1966) and by Baldwin and Baldwin (1973). With increasing age the ritualization becomes weaker, and at the beginning of subadulthood, play takes more and more the form of actual fighting.

Rhesus infants who did not pass through the stage of play because they had been reared in complete isolation, or only with their own mother or an artificial mother substitute, behave more aggressively in later peer contacts (Harlow and Harlow 1965, Mason 1963, Mitchell 1968), while sexual behavior and grooming are less frequent (Mason 1965b, Spencer-Booth 1969). Baldwin and Baldwin (1973) observed two squirrel monkey troops in which juveniles did not show any play behavior. At the same time, subadult males did not engage in consort relationships with females, sexual behavior of adults consisted of minimum sequences, and subadult males did not groom among each other. The bonding-stage model would suggest that in all these cases play is the lowest bonding stage in immature peer contacts, and that its omission impedes the further development of social relationships.

While play frequencies agreed with the dyadic rules, deviations were found in the grooming frequencies (fig. 17). They concerned mainly the grooming interactions of immature females. To explain them, three hypothetical triadic effects had to be postulated: (1) Interactions between mature males and immature females suppress the females' grooming interactions with immatures. (2) Harem leaders suppress interactions between immature females and subadult males. (3) Interactions between mature males and immature females suppress grooming between mature and immature males. These triadic effects have two common features. First, they all involve a male-female relationship that reduces the interactions of males with the pair. This triadic rule has already been verified for adult animals (Kummer et al. 1974). Second, they all seem to be related to family forming. The second triadic effect delays the formation of initial units. The first and third effects reduce interactions of initial units with outsiders. The three triadic effects will be further investigated in the following chapters.

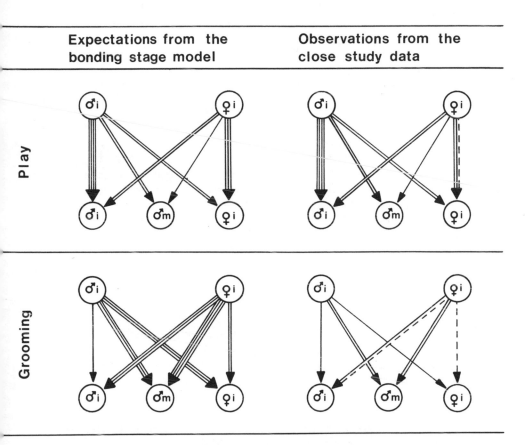

Figure 17. Graphic representation of the dyadic rules in social interactions of immature animals. The figure summarizes the results of table 15. Within each graph the upper circles represent actors, the lower circles partners of interactions; i = immature, m = mature. Within each graph an arrow of a given thickness indicates a frequency which is significantly higher than any thinner arrow ($p < 0.05$). Arrows with dotted lines represent intermediate frequencies. The graphs illustrate that play interactions agree with the dyadic rules, whereas grooming interactions involving immature females seem to be suppressed.

5 The Forming of the Initial Unit

The preceding chapter was not easy reading, because the investigation of the social relationships in immatures emphasized figures rather than animals. In this chapter attention will be shifted from immatures to mature males, from sex-age classes to individuals, and from the whole Cone Rock troop to band I. The subject of this chapter is the development of the initial unit, which consists of a young adult male and one or more juvenile females and which represents an ontogenetic precursor of the fully developed one-male unit (Kummer 1968a). The transition from sex-age classes to individuals permits the introduction of field records that serve to illustrate the variety and complexity of family forming. A focus on band I is advisable because in band I all but one mature males involved in family forming belonged to the close-study sample.

Throughout the next chapters I shall use the following terminology to describe the social status of mature males. A mature male who possesses at least one adult female is a *harem leader*. A mature male who possesses one or more juvenile but no adult females is an *initial leader*. A mature male without any females is a *bachelor*. A mature male, who is associated with a one-male unit is a *follower*, regardless of whether he is also an initial leader. Figure 18 presents the names, age classes, and family status of the mature males in band I. The reader is invited to turn occasionally to this figure during the reading of the next chapters. Table 19 gives the compositions of the one-male and initial units in band I.

In band I all mature males were either harem leaders or followers. The followers were either initial leaders or bachelors. The leader-follower associations were qualitatively evident, but they were not systematically quantified. Characteristic features of the follower status were: *(a)* The follower used to leave and arrive at Cone Rock together with his harem leader. Table 20 gives the number of instances the followers were recorded as arriving or leaving within the moving parties of one-male units. *(b)* When not near the band's periphery or near play groups, the follower used to stay in the vicinity of his one-male unit and to interact with its members. Some figures concerning this aspect will be presented in this and the following chapters. *(c)* The followers used to sleep in the vicinity of their one-male unit. *(d)* Most followers morphologically resembled their harem leaders. This aspect will be presented in chapter 7.

90

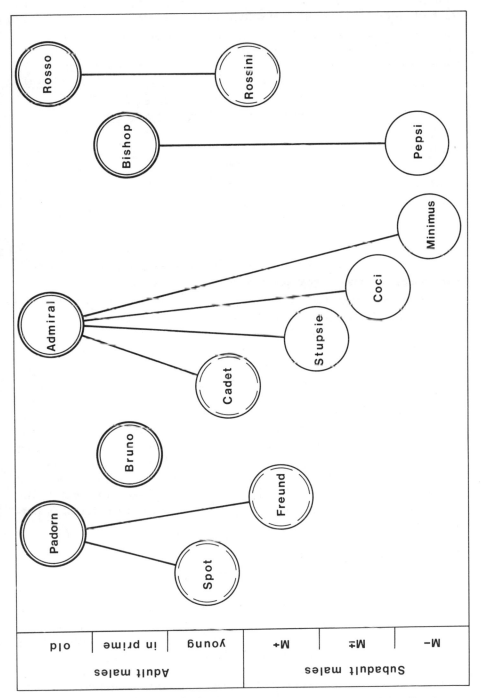

Figure 18. Names, age classes, and family status of the mature males in band I in 1971/72. The double circles indicate harem leaders; double circles with dotted lines represent initial leaders and simple circles bachelors. Lines that link males represent leader-follower associations. For definitions see text.

Table 19. The Composition of the One-Male and Initial Units of Band I in April 1972

Harem and initial leaders	Males					Females					Total
	sa	3	2	1+B	N	a	3	2	1+B	N	
Padorn	1	–	1	–	–	2	–	1	2	–	8
Bruno	–	–	–	–	–	2	–	–	2	–	5
Spot	–	–	–	–	–	–	1	–	–	–	2
Admiral	3	–	2	2	2	9	–	–	7	–	26
Cadet	–	–	–	–	–	–	–	2	–	–	3
Rosso	–	–	1	3	–	4	–	–	1	1	11
Bishop	1	–	–	–	–	2	–	–	1	–	5
Rossini	–	–	–	–	–	–	–	1	–	–	2

For abbreviations of sex-age classes see table 1.

A. The Development of the Initial Bond in Band I

The interactions between followers and immature females in band I seemed to represent different stages of a single process. Figure 19 indicates a positive correlation between the age of the followers and the age of the immature females with whom the followers mainly interacted. The process was characterized by decreasing maternal behavior and increasing sexual behavior and grooming between the followers and the immature females. This agrees with the description given by Kummer (1968a). This simple change of behavior, however, conceals a complex process that has to be viewed in more detail. The whole process of forming an initial unit can be divided into two phases: the *onset* of the initial bond and the *separation* of the initial unit from the female's parental unit.

1. The Onset of the Initial Bond

The first phase of forming an initial unit brings together the prospective members of an initial unit and gradually strengthens the initial bond. In band I the young followers Minimus, Pepsi, Coci, and Stupsie were in this onset phase in 1971/72. Minimus is not taken into account here because he did not belong to the close-study sample. Figure 19 shows that Pepsi, Coci, and Stupsie interacted mainly with infant females. The interactions consisted of brief sequences of maternal behavior and mutual grooming. The context of these interactions has been described by Kummer (1968a, p. 61) as kidnapping. As far as the group membership of the females could be determined, more than 80 percent

Table 20. Leader-Follower Associations in Band I

		Harem leaders				
		Padorn	Bruno	Admiral	Rosso	Bishop
Followers	Spot	16 8	0 0	0 0	0 0	0 0
	Freund	45 10	0 0	1 0	0 0	0 0
	Cadet	3 3	0 0	19 3	0 0	0 0
	Stupsie	5 2	0 1	24 5	4 0	2 0
	Coci	1 0	1 0	23 8	0 0	0 0
	Minimus	0 0	0 0	11 3	0 0	0 0
	Rossini	1 0	0 0	0 0	45 8	1 0
	Pepsi	3 1	1 0	1 1	35 1	50 4

The table gives the number of instances a follower was recorded to move or to stay within the party of a given one-male unit. The upper figure refers to the records at the sleeping cliff, the lower to the observations on the foraging march.

of Pepsi's, Coci's, and Stupsie's interactions with immature females occurred with females of the one-male units of which the males were followers.

Eighty-six percent of *Pepsi*'s interactions with immature females were addressed to Prima, the only immature animal in the one-male unit of Bishop, of which Pepsi was a follower. Pepsi regularly mothered and groomed Prima. These interactions represented 96 percent of Prima's interactions with mature males. Interactions between Pepsi and Prima occurred sporadically. Prima still belonged fully to her parental unit. She interacted five times more frequently with her mother than with Pepsi; at night she slept near her mother, never near Pepsi. At the beginning of 1972 a female infant was born in the one-male unit of Rosso. This female, Blondy, aroused the interest of Pepsi. As soon as Blondy started to move away from her mother, Pepsi attempted to approach her. On several occasions he succeeded in mothering Blondy; twice he mounted her and once he groomed her. The relationship between Pepsi and Blondy lasted for three months. During this period Pepsi moved with and stayed near the one-male unit of Rosso, Blondy's parental unit (see tab. 20); he neglected Prima and did not interact with her. After that Pepsi was again regularly found near the one-male unit of Bishop, where he resumed his interactions with Prima.

Coci and *Stupsie* were followers of Admiral. Both of them interacted

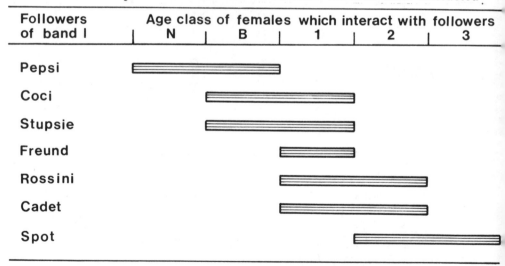

Figure 19. Age class of immature females that interacted with the followers of band I in 1971/72. The close-study followers of band I are ranked according to increasing age. Interactions between followers and immature females include maternal, sexual behavior, and grooming. The graph presents those females who together account for more than 90 percent of the followers' interactions with immature females. Interaction frequencies are summarized in table 21.

ble 21. Family Forming in Mature Close-Study Males: Behavioral
         Frequencies for Interactions with Immature Females

age of mily rming	Age class	Band	Individuals	Mothering	Mutual grooming	Followed by imm. females	Checking on following response
formation of initial unit	M−	III	Garu	0	0	0	0
	M−	III	Jodok	0.1	7.3	0	0
	M+	III	Pirat	0	0	0.1	0
	a⁻	III	Steward	0	0	0.2	0
			Mean	0	1.8	0.1	0
bachelors in onset phase	M−	II	Jaso	0.4	2.3	0.9	0
	M−	I	Pepsi	0.3	1.7	0.1	0
	M−	I	Coci	0.4	8.3	1.7	0
	M+	II	Husar	2.4	0.3	0.1	0.1
	M∓	I	Stupsie	0.2	9.2	0.3	0.1
			Mean	0.7	4.4	0.6	0
initial leaders	M+	I	Freund	0.1	3.0	1.4	0.4
	M∓	I	Rossini	0	27.0	1.9	0.8
	a	I	Cadet	0	36.5·	0.8	0
	a	I	Spot	0	19.6	1.5	1.1
			Mean	0	21.5	1.4	0.6

ιe close-study males are classified according to their stage of family
»rming (see text). Frequencies (in percents) are calculated for the whole
»servation periods of each male, except for Pirat and Steward, whose
»rupt takeover of immature females are not taken into consideration since
ιe table intends to illustrate that no interactions with immature females
·ecede the formation of initial units in band III. Testing the frequencies
·ross male and behavior categories yields significant differences against
ιe estimation of error (F_e = 11.85, p < 0.01), but not against the
·atistical interactions (F_i = 2.27), p > 0.05), because the between-
ιtegory differences are particularly great for grooming.

exclusively with immature females of Admiral's one-male unit. Certainly the females were not always the same individuals, because the bachelors interacted within the same period of observation with females N and B. But since the females were not marked we could not discover any preferences for particular individuals. Coci's and Stupsie's interactions with immature females occurred in short bouts; the females remained within their parental unit.

In band II, three subadult males who belonged to the close-study sample showed the same behavior toward immature females as did Pepsi, Coci, and Stupsie (see tab. 21).

From these limited observations three features may be characteristic of the onset phase of the initial bond: (1) the bachelors interact mainly with females who belong to the one-male unit of which the bachelors are followers; (2) interactions between bachelor and prospective initial female characteristically include maternal behavior, and (3) the interactions occur in short bouts of about 10 to 30 minutes; the female still belongs fully to her parental unit.

2. The Separation from the Female's Parental Unit

The second phase of forming an initial unit separates the initial female from her parental unit and leads to the autonomy of the new initial unit. In band I Freund, Rossini, Cadet, and Spot represented various stages of this process.

Freund appeared to be at the beginning of separating his initial female from her parental unit. He was a follower of Padorn and sustained a relationship with Freundin, a female who was at the end of age class 1 and who belonged to the one-male unit of Padorn. Freund interacted with no other immature female. He showed vigorous and obstinate efforts to separate Freundin from Padorn's one-male unit and used a variety of techniques to reach his goal. When the band arrived at Cone Rock in the evening, Freund tried repeatedly to place himself between Freundin and her parental unit. While Padorn and his one-male unit climbed the cliff, Freund prevented Freundin from joining her one-male unit by skillful interposition:

> The harem unit of Padorn slowly climbs the rock. Freundin, who sits half a meter behind Freund, intends to follow her parental unit while glancing at Freund. Freund immediately gets up. Freundin sits down; so does Freund. This sequence is repeated. Then Freund slowly climbs the rock, repeatedly looking back at the female who follows him. As Freund sits down, the female tries to pass him. Freund immediately gets up and walks in front of Freundin. Three meters in front of Padorn's one-male unit, which has by now reached its usual place, Freund sits down. Freundin approaches and grooms him. Six times she stops grooming and attempts to pass Freund in order to join her parental unit. But each time Freund blocks access to Padorn's unit by interposition. Twice Freundin withdraws in the opposite direction and tries to reach her unit on a detour. Freund follows attentively and barricades the access to Padorn's one-male unit while yawning. The whole sequence lasts about 20 minutes.

Freund was not always successful. Sometimes Freundin groomed him so peacefully that his attention decreased; then she suddenly ran away and attained her one-male unit before Freund could reach her again.

The following sequence, which occurred twice within one week, appeared as if Freund was testing the female's developing readiness to remain near him:

> Freund and Freundin have climbed the rock, Freund preceding the female. As Freund approaches the one-male unit of Padorn, he suddenly runs away from Freundin and sits above Padorn's unit, so that the female may approach either him or her parental unit. She joins her unit.

> One week later, Freund and Freundin again climb up to the place where the one-male unit of Padorn has settled. Freund stops and invites the female to groom him. As the one-male unit of Padorn moves, Freundin stops grooming. Freund remains seated, then moves a little toward Padorn's unit and stops again. Freundin follows and passes Freund who looks at her. The female sits and scratches. Freund climbs in front of her and walks with her on a detour toward the one-male unit of Padorn. Again he places himself quickly aside, so that the way to Padorn's unit becomes free to Freundin. But this time the female approaches Freund and grooms him. In the course of the grooming interaction, Freund walks away in small shifts until he is about 8 meters from Padorn. Freundin follows him each time.

Freund showed another characteristic behavior that struck our attention. Several times we saw him standing a few meters from Padorn, with stretched arms and legs, the head bent sideways and the tail raised. He looked as if he were being groomed by some animal, but when he moved after several minutes, we were sure that in fact nobody had groomed him. Freund's behavior looked like an invitation to grooming from a distance, although we never saw the invitation accepted by Freundin.

If Freund failed to keep the female in his vicinity, he mostly did not engage in any other social interaction, but climbed down to his sleeping site on a small edge of a steep cliff. He was then the only animal to occupy his sleeping ledge that early in the evening.

After about four months we noticed some effects of Freund's efforts. Freundin attempted less often to run away from Freund, and Freund appeared less vigorous in his interactions with the female. When the band arrived at Cone Rock, it was often sufficient for him to sit down while the one-male unit of Padorn passed on, and the female would remain with Freund. At night, however, Freundin regularly slept within the one-male unit of Padorn; a few times she showed an intention to approach Freund on his sleeping ledge, but then returned to her parental unit. During this period Freund attempted several times to copulate with the female who still had not started to cycle sexually. Shortly before the initial bond between Freund and Freundin would have reached full autonomy, Freundin disappeared (see section 6A)

That Freund's efforts to separate his initial female from her parental unit was probably a continuation of what we have described as the onset of the initial bond was suggested by Pepsi. In 1974, during my short field visit, Pepsi clearly attempted to separate Prima from the one-male unit of Bishop with techniques similar to those of Freund. By 1975 Pepsi and Prima had become an autonomous initial unit (Sigg and Stolba, personal communication).

The same process of separating the initial female from her parental unit was accomplished by *Rossini*. Rossini was a follower of Rosso and interacted most frequently with Rosa, the female 2 in the one-male unit of Rosso. These interactions represented 91 percent of Rossini's interactions with immature females. At the beginning of our observations Rosa still belonged to the one-male unit of Rosso. The first record of Rossini, in June 1971, already showed his tendency to leave the one-male unit with the juvenile female:

> Rosa grooms Rossini 2 meters from her one-male unit. Rosso moves with his harem, and Rosa follows immediately. She then stops between her leader and Rossini, sits down, scratches, and looks at Rossini, who approaches her. Rossini slowly passes the one-male unit of Rosso, followed by Rosa. He stops, Rosa passes him, stops also and looks at him. Rossini yawns, moves on, and stops again. Rosa follows and grooms Rossini at 5 meters from Rosso. Rossini moves again, repeatedly looking back at the female. Rosa follows slowly. Twenty meters from Rosso Rossini climbs a small acacia. Rosa stays behind; she alternatively walks a few steps, then stops and feeds. Six meters from Rossini she sits down and feeds. Then she returns to her one-male unit. Rossini remains in the tree.

Four days later a similar sequence occurred. Rossini walked away and climbed a tree 30 meters from Rosso. Rosa followed hesitantly and with frequent feeding stops. After more than 10 minutes she finally climbed the tree; she sat near Rossini and groomed him.

Apart from two relapses which occurred in September and in October 1971, the bond between Rossini and Rosa gradually became closer. By the end of 1971 the pair appeared as an autonomous unit: in the evening they climbed the rock together while Rosso and his harem remained in the dry riverbed; on the rock Rosa remained near Rossini and returned only occasionally to her parental unit. In this phase of relative stability Rossini tried repeatedly to copulate with Rosa, who had not yet had her first estrus. According to our qualitative impression the main difference between the initial bonds of Freund and Rossini seemed to lie in the behavior of the females: while Freundin appeared unwilling to follow Freund and attempted to return to her parental unit whenever possible, Rosa seemed ready to remain in the vicinity of Rossini.

Cadet's initial unit consisted of three juvenile females. He interacted with no other immature females. Cadet acquired his females not one after the other, but more or less together in one process. He was a follower of Admiral, whose one-male unit consisted of 9 adult and 7 immature females. The size of this one-male unit may explain the size of Cadet's initial unit. In summer 1971 the initial bond between Cadet and the females did not yet constitute an autonomous initial unit; the juvenile females of Cadet were often found in the one-male unit of Admiral, where they groomed with adult females. From the beginning of our observations, however, Cadet made no efforts to keep the females near him. Whenever he walked away, the females followed him, and Cadet did not even check on the following response by looking back at the females (see tab. 21). Only twice was Cadet observed to show aggressive herding toward his females:

> Cadet chases one of his females and bites her on the neck. The female runs screaming into the one-male unit of Admiral. Cadet follows her, stops in front of Admiral's one-male unit, pumps cheeks, and threatens the female. Then he sits down and wipes his muzzle. The juvenile female approaches Cadet, who embraces her while yawning. She grooms Cadet.

The most advanced stage in the development of the initial unit was demonstrated by *Spot*. Spot was the oldest among the followers of the band. Together with Freund he was associated with the one-male unit of Padorn. Elizabeth, his female 2–3, was permanently found in the vicinity of Spot. Only at the beginning of our observations, in summer 1971, was Elizabeth occasionally observed in the one-male unit of Padorn. Spot never showed any maternal behavior or aggressive herding toward Elizabeth.

The behavior of the older followers in band I suggested that the separation of the initial bond from the female's parental unit was a consequence both of the follower's activity and of the female's readiness to follow him. From the follower's side the separation was not marked by a spectacular takeover with fighting and aggressive herding, but by subtle and steady tactics of interposing and leading.

3. Band-Specific Differences in Forming the Initial Unit

Of the 13 mature males who were sampled in the close study (see tab. 2), 7 belonged to band I. Two males were from band II and 4 males from band III. Did these males from bands II and III reveal the same gradual process of forming an initial unit that I have described for band I?

The two males of band II showed similar behavior toward immature females to that of their age mates in band I—Pepsi, Coci, and Stupsie. A third close-study male of band II, who still belonged to age class 3, behaved also in the way that I considered typical for the onset phase. Whether the interactions of the three males with immature females finally led to the establishment of initial units remains unknown because in 1973, band II had left Cone Rock (see section 3C), and by 1974 I could no longer identify the three males.

The close-study males of band III seemed to behave differently. Two of the four males, Garu and Pirat, were followers of one-male units with immature females; but none of them appeared to be interested in immature females. Pirat was a follower of a one-male unit with nine adult and six immature females; from June 1971 until May 1972 he was found mostly in the vicinity of this one-male unit. We never saw him interact with any immature female. Between two observation sessions seven days apart Pirat suddenly acquired a female 2. We do not know from where he had taken the female except that she did not come from the large one-male unit of which Pirat was a follower. After the acquisition of the female, Pirat preferentially stayed at the band's periphery and seemed not to follow the large one-male unit any longer. The other two close-study males, Jodok and Steward, seemed not to be followers. They were regularly found near the band's periphery in a party of bachelors who formed a subunit within band III. Jodok was observed once to groom a female B. Steward did not interact with immature females, but, after a two-weeks interruption of observation, in July 1972, Steward was found to have acquired a female 1.

These observations suggest that the bachelors of band III, whether followers or not, did not interact regularly with immature females. The forming of initial units of two males occurred as a short and sudden takeover, in contrast to the gradual edging loose in bands I and II. The mature close-study males, therefore, can be divided into three classes, which are presented in table 21: the bachelors of band III, who hardly interacted with immature females; the bachelors of bands I and II, who were in the onset phase of the gradual development of initial units and whose interactions with immature females characteristically included maternal behavior; finally, the initial leaders of band I, who were separating or had separated their initial females from their one-male units. These classes will be used in the following sections to relate some behavioral frequencies to the stages of family forming.

B. Interactions between Initial Bonds and Harem Leaders

We return to the gradual type of initial bond. Evidently, the process of forming an initial unit did not take place in a social vacuum. Earlier, deviations from the dyadic rules suggested the hypothesis that harem leaders have an inhibitive effect on the development of initial bonds. Therefore, we are particularly interested in studying the response of the juvenile female's leader and probable father to the forming of the new family unit. Since harem leaders did not belong to the close-study sample, I cannot directly and quantitatively demonstrate this inhibitive effect. Accordingly, this section contains only qualitative hints, which make plausible the harem leader's effect on the developing initial unit.

1. Overt Interference of Harem Leaders with Developing Initial Units

Harem leaders were observed only a few times to intervene directly in interactions between immature females and bachelors. Occasionally harem leaders chased away their followers, but mostly we had not observed the preceding events. The following field records illustrate interventions of harem leaders:

Pepsi sits near the one-male unit of his leader, Bishop. Prima, the female B of the unit, approaches Pepsi. Bishop runs after her, grabs the female, and carries her on his back to his one-male unit. Prima attempts again to approach Pepsi, and Bishop again runs after her and grabs her. While Bishop remains seated with Prima, Pepsi withdraws from Bishop and sits near the adult females of Bishop.

Freund was chased away several times by his leader Padorn. Since Freund's social activity was exclusively aimed at Freundin, it is probable that the interactions between Freund and Freundin provoked Padorn's intervention. But Padorn could be more subtle in keeping Freundin within his one-male unit and in bringing a sufficient distance between his harem and his follower:

Freund sits 1.5 meters from Padorn's one-male unit. He looks at Freundin and yawns and scratches frequently. Freundin, who sits within the one-male unit, scratches, then grooms herself. Once she approaches Freund; Padorn yawns and Freundin returns immediately to the one-male unit. The tense situation with frequent conflict behavior goes on, until after some minutes Padorn leaves the sleeping site of his unit and moves his unit to an uncommon place. Freundin follows and then grooms her mother at the new place. Freund remains seated and starts self-grooming.

The development of Rossini's initial unit was thrown back twice by interventions of Rosso during the period of separating Rosa from her parental unit: when Rosa was trapped for marking, Rossini demonstrated his claim of possession by staying near the trap and vigorously threatening us. We do not know whether this demonstrative behavior had any influence on Rosso, but when we released Rosa in the evening, Rosso and Rossini chased each other and within a few seconds the animals of Rosso's unit were out of sight. When we found them again, Rosa was near Rosso and groomed him briefly, while Rossini sat in a distance of about 20 meters. During the following weeks Rosa remained within the one-male unit of Rosso and no interactions between her and Rossini were observed. It was one month after the release of Rosa that we again recorded an interaction between Rossini and Rosa. From then on the initial bond became gradually an autonomous unit which sometimes stayed 20 meters from Rosso's one-male unit. One month after the first new interaction between Rossini and Rosa we observed the following sequence:

> As band I arrives at Cone Rock, Rosso and Rossini have bleeding wounds on their faces. Rossini runs at Rosa, who has walked away from her parental unit, and crouches over her. Immediately Rosso lunges at his follower. There is a short but vigorous fight; then Rosso chases his follower away. Rosso climbs the rock with his one-male unit, while he and Rossini continue to pump cheeks and yawn. Rosa is within her parental unit. Repeatedly she shows clear intentions to approach Rossini, but returns to a place behind Rosso as soon as the males start to threaten each other.

It took about one more month for the initial bond between Rossini and Rosa to recover. No further intervention of Rosso was observed during the remaining eight months of the study.

We gained the qualitative impression that harem leaders had a possessive claim on immature females, but not on immature males. As a consequence they seemed to intervene preferentially when subadult males attempted to interact with immature females. We collected 13 nonrandom samples that contain an adult male's interference with an interaction between nonadult animals. In 8 protocols a subadult male and in 5 protocols a juvenile attempted to approach another animal and were prevented by the adult male's intervention. Ten of the 13 approached animals were immature females, all of them in age classes N to 1. By our trapping we made an involuntary experiment—the result of which is compatible with the harem leader's possessive claim on immature females: when we had trapped immature males, mainly subadult males stayed around the traps. The harem leaders would leave with their one-male units, even when enough food and water was available near the sleeping cliff; the subadult males near the traps would follow the departing troop as the last animals. When immature females were trapped, the harem leaders remained near the traps and defended the trapped females.

2. Inhibitive Effect of Harem Leaders on Developing Initial Units

The observation of the rare interventions of harem leaders would not have led us to the hypothesis that harem leaders might negatively affect the development of initial units. The hypothesis was mainly suggested by the behavior of followers. This behavior resembled the behavior of inhibited rivals in the experiments of Kummer et al. (1974). In a triadic situation with a male-female pair and a male rival Kummer et al. found the following features to be typical: the inhibited rival would neither approach nor interact with the pair; instead he would stay rigid, look away from the pair, and exhibit frequent conflict behavior (fiddling on the ground, wiping the muzzle, scratching, and self-grooming); the owner of the female would typically notify the rival rather than vice versa. Spatial proximity and one casual interaction in the pair seemed to be sufficient to cause the rival's inhibition.

The following records and figures from the close study illustrate the similarity between the behavior of bachelors toward immature females and the behavior of inhibited rivals:

If harem leaders have a possessive claim on immature females, but not on immature males, we should expect that bachelors come more easily into contact with immature males than with immature females. This is contrary to the expectation from the dyadic rules, which would predict a higher compatibility for interactions between subadult males and immature females. In fact, the close-study bachelors groomed significantly more often with immature males than with immature females (Wilcoxon matched-pairs signed-ranks test, $P < 0.05$). Moreover, subadult and immature males had in their interactions a ratio for bodily contact/approaching of 0.66 (sample size = 273 approaches), whereas between bachelors and immature females the ratio was 0.41 (sample size = 151 approaches). The difference is significant ($X^2 = 7.25$, $p < 0.01$). This means that an approach between a bachelor and an immature female had a smaller probability of leading to a tactile interaction than an approach between a bachelor and an immature male. This result agrees with our observation that bachelors who were near a play group with immature females appeared very hesitant to interact with the females. The following field record illustrates the behavior of a bachelor in the vicinity of a play group with immature females:

Stupsie sits in front of a group including female infants. A juvenile male passes and sits nearby. Stupsie yawns. He looks at the infants, scratches, yawns, scratches and yawns again. A female infant approaches him. Stupsie grabs her, pumps cheeks, and yawns. The infant leaves him; Stupsie scratches, masturbates, grooms himself, and yawns. Again a female infant approaches him. Stupsie scratches, stands up and yawns. The female walks away. Stupsie looks at the female, sits down with an erection. The female group moves. Stupsie follows, remains standing, yawns twice, and sits down. The group moves again. Stupsie follows, sits with an erection, yawns, wipes his muzzle, grooms himself, scratches twice, constantly looking at the females; he scratches, wipes his muzzle, follows the group which has moved, stands, looks at the females, sits down, scratches, yawns. This sequence lasted 10 minutes.

By 1972 Rossini had established an autonomous initial unit with Rosa. In March and April 1972 Rossini took interest in Blondy, a female infant in Rosso's one-male unit. He hardly ever engaged in tactile interactions with the female infant, but he performed the characteristic hesitant behavior of a bachelor toward an immature female; in addition, Rossini neglected his initial female, Rosa, as did Pepsi with his prospective initial female, Prima, in the same situation.

Rossini sits near the one-male unit of Rosso and looks continuously at Blondy. Rossini's initial female, Rosa, is out of sight. A few minutes later Rossini is groomed by a male 1 from Rosso's one-male unit. After five more minutes Rosa appears and sits 2 meters from Rossini, who ignores her and continues grooming with the juvenile male. Then Rossini walks away, sits 1 meter from Rosa, but looks constantly at Blondy, who is in a small play group. He wipes his muzzle, yawns, shakes his body, and scratches. He pays no attention to Rosa but starts grooming with another juvenile male from Rosso's one-male unit.

Rossini follows the one-male unit of Rosso which moves on the rock. Eight meters from the one-male unit he sits down and feeds. He ignores Rosa who sits 3 meters behind him. He climbs a small acacia and looks at Blondy while Rosa joins her parental unit. When the band leaves for the foraging march, the one-male unit of Rosso passes Rossini who still looks at Blondy. At the rear of the one-male unit Rosa walks and stops in front of Rossini, looking at him. Rossini pays no attention to his initial female, who follows her parental unit.

If conflict behavior in bachelors is increased because they are inhibited from interaction with immature females, we should expect that it decreases in frequency as soon as the followers have overcome the inhibition and established autonomous initial units. This assumption is confirmed by table 22, which classifies the mature males according to their status of family forming. The table suggests, in addition, that yawning and wiping the muzzle might be typical for bachelors in the onset phase (Jaso, Pepsi, Coci, Husar, and Stupsie) and, together with table 21, that self-grooming might be negatively correlated with heterosexual grooming interactions. The following figures also agree with the assumption that reduced access to immature females coincides with increased conflict behavior in followers. During the period in which Pepsi attempted to interact with the female infant Blondy, his frequency of conflict behavior increased from 13 to 21 percent ($X^2 = 17.22$, $p < 0.001$); and during the period in which Rossini's access to Rosa was blocked because of Rosso's interventions, Rossini's conflict behavior increased from 7 to 13 percent ($X^2 = 12.58$, $p < 0.001$).

How can we know that the harem leaders of immature females were

Table 22. Family-Forming in the Mature Close-Study Males: Frequencies
           of Conflict Behavior

Stage of family forming	Age class	Band	Individuals	Self-grooming	Yawning	Scratching	Wiping muzzle
No gradual formation of initial unit	M−	III	Garu	4.5	1.1	7.9	0.2
	M−	III	Jodok	6.2	1.9	9.6	0.3
	M+	III	Pirat	5.3	1.6	8.0	0.3
	a⁻	III	Steward	4.2	2.9	9.1	0.5
			Mean	5.1	1.9	8.7	0.3
Bachelors in onset phase	M−	II	Jaso	4.2	5.1	13.8	1.3
	M−	I	Pepsi	6.2	0.7	8.7	0.4
	M−	I	Coci	2.1	1.1	5.5	0.7
	M+	II	Husar	1.3	4.5	6.7	0.3
	M+	I	Stupsie	5.1	2.5	8.4	0.6
			Mean	3.8	2.8	8.6	0.7
Initial leaders	M+	I	Freund	3.0	1.0	1.7	0.3
	M+	I	Rossini	0.8	1.5	4.8	0.3
	a	I	Cadet	4.2	0.4	7.2	0.4
	a	I	Spot	1.3	0.7	5.7	0.3
			Mean	2.3	0.9	4.9	0.3

The close-study males are classified according to their stage of family
forming as in table 21.   Testing the frequencies across male and
behavior categories yield significant differences against the
estimation of error (F_e = 8.46, p < 0.01) and nearly significant
against the statistical interactions (F_i = 4.79).

the cause of the followers' inhibited behavior? Until the question is
answered experimentally, we have to refer to qualitative hints: casual
interactions between harem leaders and infants did occur. It seemed that
harem leaders differentiated between male and female infants as they did
between trapped male and female immatures. Out of 8 marked female
infants N to B, 5 were observed to be mounted at least once by their
leaders, whereas out of 7 male infants only 1 male was mounted by his
leader. As to spatial proximity, which also seems to contribute to inhibition (Kummer et al. 1974), females N to 1 tended to stay nearer to their
harem leaders than did males N to 1: the females interacted more frequently with their mothers within one-male units (see section 4E), while
males were found more often in play groups (see section 3D). Kummer
(1968a, pp. 91, 93) found in his results on the ontogeny of grouping
tendencies that one-year-old males joined predominantly male play
groups, whereas one-year-old females showed a significant preference
for adult males as second neighbors. All these hints agree with the view

that a harem leader and an immature female as a "pair" are more likely to inhibit a subadult male than are a harem leader and immature male. The following observations also suggest the harem leaders' negative impact on interactions between subadult males and immature females:

Subadult males usually did not follow or approach immature females within the harem's core area, which lay around the harem leader. Whenever Freundin succeeded to escape from Freund (see above), Freund would chase after her, but stop immediately when she had reached her parental unit. A similar behavior of Cadet is illustrated in the field record on page 99. Rossini, as well, never attempted to approach Rosa when she stayed in the one-male unit of Rosso (see above).

Sometimes interactions between bachelors and immature females became possible when both animals were out of the leader's sight or when the harem leader's attention was absorbed in other activities.

> Pepsi sits near the one-male unit of Rosso. When Blondy leaves her parental unit, Pepsi follows her. He precedes the female infant and leads her to a place 2 meters behind Rosso's back. There Pepsi hastily mounts Blondy. Rosso notices the incident and yawns. Pepsi withdraws immediately.

> Pepsi lies sphinxlike on his belly and gazes at Blondy, who is in a play group. Then he stands up, looks at Blondy, and yawns. In the moment when Rosso moves away with his harem, Pepsi approaches Blondy, sits, looks at her, wipes his muzzle, and scratches. Then he grabs and embraces her. Blondy walks away. Pepsi approaches her again, grabs her, sits, and yawns, while Blondy follows her walking parental unit. Pepsi remains seated with his elbows within the knees.

Three subadult males were observed once each to grab and carry away an immature female in the moment when the female's harem leader was engaged in a fight with another male. That the absorbed attention of a harem leader increases the probability for interactions between subadult males and immature females will be illustrated for Pepsi, Minimus, and Stupsie in the next chapter.

In only one case was the inhibited male observed to direct aggressive behavior toward the female's leader:

> 12 meters from Rosso Rossini lies sphinxlike on his belly and gazes continuously at Blondy. He nervously moves his tail and twice scratches erratically on his anal field. Then he suddenly stands up, stares, and lifts brows towards Rosso, and again lies on his belly.

To lie sphinxlike on one's belly was an uncommon baboon behavior and was observed only in the context of inhibited behavior toward immature females; when merely relaxing, the baboons lay on their sides.

Figures 20a, b, c, and d. Behaviors of followers that are interpreted as social inhibition toward immature females: (a) Lying sphinx-like on the belly; (b) sitting with bowed back, bent head, and elbows within the knees; (c) a follower sits near a play group including immature females and yawns; (d) a follower is approached by a juvenile female; he turns away from the female, whereas the harem leader runs after her.

To lie on the belly appeared to be a form of staying rigid, similar to sitting with bowed back, bent head, and elbows within the knees, a posture that occurred also in the experimental situation of Kummer et al. (1974).

Figure 20 illustrates some typical behavior of bachelors toward immature females, behaviors that resemble those of inhibited rivals. This inhibition I postulate for bachelors seems not to be complete; it reduces interactions between bachelors and immature females, but does not suppress them entirely. It appears to be weaker than the inhibition of rivals toward an adult male-female pair. Two factors might account for the incomplete inhibition: (1) Within one-male units immature females interacted less frequently with harem leaders than did adult females. In the broad sample, which was our only sample including harem leaders as well as adult females, adult males groomed nine times more often with adult than with immature females (see also Kummer 1968a, fig. 40). The father-daughter bond could, therefore, represent a less conspicuous and hence less effective pair gestalt than the harem bond. (2) The capability of becoming inhibited may be less developed in subadult than in adult males. This theoretical possibility, however, remains unsupported. The only subadult male in the experiments of Kummer et al. (1974) did not show a weaker inhibition than the adult males.

C. Interactions of Followers with Immatures

The investigation of the dyadic rules in chapter 4 has led to the hypothesis that grooming between mature males and immature females suppresses grooming between mature and immature males (see section 4D); furthermore, the troop differences in the broad sample data of Cone Rock and White Rock suggested that play among males decreases in the age that precedes the establishment of the initial unit (see section 3E). These hypotheses are tested now within the close-study sample of mature males. Table 23 compares frequencies of interactions with male partners for bachelors of band III, bachelors of band I and II, and initial leaders of band I. The table contains interactions with all, not only with immature, male partners for the following reason: chapter 5 will suggest that grooming between young subadult and immature males persists in ontogeny and appears later as interactions between young adults and young subadults. If we test the above mentioned hypotheses on all mature close-study males, who range from young subadults to young adults, we have to take into account all male members. Tables 21 and 23 show that increasing interactions with immature females tend to be negatively correlated to play, mothering, and grooming interactions with male partners. Grooming with immature females is negatively cor-

Table 23. Family Forming in the Mature Close-Study Males: Behavioral Frequencies for Interactions with Male Partners

Stage of family forming	Age class	Band	Individuals	Play	Mothering	Grooming	Notifying
No gradual formation of initial unit	M–	III	Garu	3.5	1.8	35.1	0.1
	M–	III	Jodok	0.2	2.1	38.1	0
	M+	III	Pirat	0.3	0.2	15.6	0
	a	III	Steward	0.5	0.7	39.5	0.2
			Mean	1.1	1.2	32.1	0.1
Bachelors in onset phase	M–	II	Jaso	0.1	0.3	11.4	0.3
	M–	I	Pepsi	0.2	0	16.5	0
	M–	I	Coci	0	0.3	27.7	0.4
	M+	II	Husar	0.1	0.1	22.7	0.4
	M+	I	Stupsie	0.5	0	19.6	0.1
			Mean	0.2	0.1	19.6	0.2
Initial leaders	M+	I	Freund	0	0	10.1	0.2
	M+	I	Rossini	0	0.4	3.4	0.6
	a	I	Cadet	0	0	11.4	1.0
	a	I	Spot	0	0	6.1	0.5
			Mean	0	0.1	7.8	0.6

The close-study males are classified according to their stage of family forming as in tables 21 and 22. Testing the frequencies across the male categories and the behaviors of play, mothering, and grooming yield significant differences against the estimation of error (F_e = 29.94, p 0.01), but not against the statistical interactions (F_i = 5.26, p > 0.05) because the between-category differences are particularly great for grooming. The differences of notifying are significant between the three male categories (F = 6.22, p < 0.05).

related to grooming with male partners (Spearman rank correlation coefficient = −0.55, p < 0.05, one-tailed). Only notifying, which is a ritualized presenting among males, is positively correlated with grooming between mature males and immature females (Spearman rank correlation coefficient = 0.66, p < 0.05). The decrease of grooming and the increase of notifying among males represents a regression of male-male relationships, if interpreted from the viewpoint of the bonding-stage model.

That an impeded access to immature females coincided with increased male-male interactions was also suggested by the behavior of Pepsi and Rossini across time. During the 3-month period in which Pepsi attempted to interact with Blondy, his frequency of grooming with male partners rose from 3 to 24 percent of intervals (X^2 = 155.11, p < 0.001). Rossini was observed 11 times to groom with male partners during the period of study; it occurred 8 times in the three months during which Rossini's access to Rosa was blocked (see section 5B) and during which Rossini was inhibited to approach Blondy; it occurred 3 times during the remaining nine months.

D. Exclusion among Followers

Several occasional observations indicated a competitive relationship among the followers of band I. Competition for immature females ranged from low intensities of rivalry manifested in notifying interactions to serious attempts to take over the initial female from another follower. A quantitative analysis of notifying is given in chapter 7. Here I shall present the occasional observations for other forms of competition:

Between June 1971 and July 1972 three attempted takeovers were observed in the initial leaders of band I. The first attempt occurred in July 1971:

Rosa has been marked and is released. Cadet rushes at her, crouches over Rosa and mounts her. Rossini runs near them. Cadet withdraws immediately.

The second attempt was observed in May 1972. This time Rossini unsuccessfully tried to take over an initial female from Cadet. The incident is presented in more detail in section 7A, because it involved notifying between the two initial leaders. The third attempt also occurred in May 1972 and was successful. Rossini took over the initial female of Spot, see section 6A, who had recently acquired an adult female.

The following observations illustrate further sequences of intense competitive behavior among followers:

Stupsie approaches Prima who is the prospective initial female of Pepsi. Pepsi, who sits nearby, runs toward Prima and grabs her. Stupsie withdraws.

Rosa is in a play group. She plays briefly with Pepsi. Rossini raises his brows, lunges at Pepsi, and chases him away.

Coci chases after a female 2 of Cadet. The female runs down the cliff, screaming. Coci stops above the female. From a distance of 8 meters Cadet appears from behind a bush. He stares at Coci and slaps the ground. Coci looks away. Cadet sits and continues feeding.

A subtle form of exclusiveness occurred between Rossini and Pepsi.

During March and April 1972 both males were interested in Blondy, the female N of Rosso's one-male unit. Both males aimed at staying in the vicinity of Blondy, but tended to exclude each other. Nine times only one of the males was found near Blondy, and twice both males were found simultaneously in the vicinity of the female infant. During this period we observed the first interactions between Rossini and Pepsi since the beginning of our study, which consisted of presenting and touching each other's genitals.

E. Interactions of Initial Females

So far the description of the developing initial unit has emphasized the behavior of the male. Now we turn to the behavior of the female. The investigation of the dyadic rules (see section 4D) and the troop differences in the broad sample data of Cone Rock and White Rock (see section 3E) have left us with the hypothesis that life in the initial unit reduces the female's interactions with immature partners.

Of the 8 juvenile females of the close-study sample 3 females were members of initial units and 5 females still belonged to the parental harem units. This permits us to compare behavioral frequencies between initial and non-initial females. The comparison is summarized in table 24.

1. Interactions with Mature Males

Initial females interacted more frequently with mature males than did noninitial females. Among the noninitial females Susi was notable for her high frequencies of sexual behavior and grooming with mature males, which were typical for initial females. The explanation is that Susi had been taken over by a harem leader (see section 6A); two weeks after the takeover she had an estrus (probably her first). Both the takeover and the estrus led to increased interactions with her new leader. Kummer (1968a, p. 93) interpreted the low frequencies of interactions between harem leaders and juvenile females with the observation that juvenile females are easily displaced by adult females when competing for access to the harem leader. This suggests the following triadic rule: interactions between adult males and adult females reduce interactions between adult males and juvenile females. The rule agrees also with Kummer's observation in captive geladas that bond forming between the adult male and the dominant female has priority over bond forming between the adult male and the subordinate female.

If it is generally valid that in a triadic situation the animal with the lowest interaction rate shows an increased frequency of conflict behavior, then we should expect that noninitial females more frequently per-

Table 24. Behavioral Frequencies of Initial and Noninitial Females

Family status	Age class	Individuals	Follow	Sex.b. with mature males	Grooming with mature males	Grooming with ad. females	Play with immature	Grooming with males	Play with immature	Grooming with females	Scratching	Self-grooming	Yawning
Initial females	1	Jacqueline	0.2	0.9	13.8	0	0	0	0.9	3.5	8.5	1.5	0
	2-3	Rosa	3.0	0.9	21.8	6.9	0.1	3.7	0.5	6.1	5.1	4.5	0.2
		Elizabeth	1.9	1.1	30.8	1.9	0	9.4	0.1	0	5.4	4.0	0.1
Noninitial females	1	Pamela	0	0.1	0	38.8	0.3	2.0	4.5	3.9	8.2	2.8	0
		Marina	0	0	0	16.7	2.0	9.0	7.7	5.0	9.0	3.6	0
	2-3	Biancha	0.3	0.7	0	23.9	3.4	21.4	1.9	6.1	6.1	5.6	0.6
		Vera	0.5	0.3	10.2	28.1	1.6	3.6	0.6	0.2	11.7	12.2	0.5
		Susi	0.4	3.4	16.0	14.4	1.4	0.7	1.7	2.3	14.2	16.3	0.7

The juvenile close-study females are classified as initial and noninitial females. Frequencies of active plus passive behavior are given with the exception of "follow," which is recorded only for the female. Frequencies across female and behavior categories yields the following results: Interactions with mature males appear to be more frequent in initial females (F_i = 2.00; F_e = 10.16, p < 0.01); grooming with adult females occurs more frequently in noninitial females (F = 11.39, p < 0.05).

Interactions with immature partners are more frequent in noninitial females (F_i = 15.00, p < 0.05; F_c = 4.95, p < 0.05); conflict behavior also appears to be more frequent in noninitial females (F_i = 12.00; F_e = 5.37, p < 0.05).

form conflict behavior than do initial females. According to the figures of table 24 this seemed to be the case in our sample of juvenile females. In particular, self-grooming seemed to decrease with increasing frequencies of heterosexual grooming, as has been suggested for bachelors and initial leaders (see tab. 22).

2. Interactions with Adult Females

As long as an initial unit was not separated from the female's parental unit the female would still interact with her mother. With the autonomy of the initial unit the relationship between adult and immature females and probably between mother and daughter was sharply reduced (see tab. 24). The process was promoted by the activity or attention of the initial leader.

Freund was in the process of separating Freundin from the harem unit of Padorn (see section 5A). Whenever Freund did not succeed in keeping Freundin near him she stayed in the harem unit of Padorn and groomed with her mother.

Rossini's initial bond with Rosa was set back twice because of the interventions of Rosso, once because Rossini took interest in the infant female, Blondy (see section 5B), and once because Rossini had acquired a new female who absorbed his attention (see section 6B). During these periods of a weakened initial bond, which covered a total time span of four months, Rosa was observed 16 times to groom with her mother. During the remaining eight months Rosa groomed only twice with her mother. Four times she was seen to approach her parental unit, but was prevented by Rossini:

> Rosa intends to approach her mother while looking at Rossini. Rossini yawns. Rosa runs at Rossini, who embraces her.

> Rosa has approached the one-male unit of Rosso. Rossini approaches her, and Rosa presents to him. Rossini crouches over Rosa, grasps her, and pulls her away from Rosso's harem.

One observation of Cadet illustrated Cadet's tendency to prevent interactions between his initial females and adult females:

> An initial female of Cadet sits near an adult female of Admiral. Cadet approaches the adult female and yawns. The adult female withdraws, and the initial female starts grooming Cadet.

The initial female of Spot, Elizabeth, once groomed, at the beginning of our study, an adult female of Padorn's one-male unit. For the rest she always followed her initial leader and never attempted to interact with the adult females of her parental unit.

3. Interactions with Immatures

Table 24 confirms that noninitial females played and groomed more frequently with immatures than did initial females. This result agrees with our hypothesis derived from the troop differences and the dyadic rules (see above).

As with interactions between initial females and adult females, the occurrence of interactions between initial females and immatures depended on the behavior and attention of the initial leader. In the period during which Rossini's access to Rosa was blocked or his attention was absorbed by other females, Rosa's frequency of interaction with immatures rose from 7.9 to 14.2 percent of intervals ($X^2 = 11.2$, $p < 0.001$). Rosa interacted freely and without hesitations, whereas her interactions with immatures were hesitant and wavering when Rossini payed attention to her:

Rosa sits in a play group and plays with a juvenile, then stops playing and looks at Rossini, who is sitting 4 meters away and has an erection. Rosa approaches Rossini, touches him, and returns to the play group, where she interacts with a male infant. Then she again approaches her initial leader, presents to him, and goes back to the play group. Rossini follows her and sits down. Rosa plays with a female B, then with a male 1, stops suddenly, bares her teeth, approaches Rossini, then walks away, and embraces a female B. A male 1 approaches Rosa. Rossini gets up and slaps the ground. Rosa immediately stares at the male 1 and also slaps the ground. Then she interacts with another juvenile male. Rossini approaches, stares at the male, wipes his muzzle, scratches, and sits down. Rosa stops playing, sits down, scratches, grooms herself, and then engages in nonsocial play. After two minutes she plays with a male 1. Rossini approaches; Rosa presents to him. Then Rossini and Rosa walk away from the play group.

Rosa sits 4 meters from Rossini, who dozes. She grooms the male 2 from Rosso's one-male unit. The male wants to groom her; she appears uneasy and looks nervously at Rossini. The juvenile male touches Rosa, scratches, and starts grooming, but stops immediately as Rosa continues to look at Rossini. Rosa utters contact grunts and scratches. The male invites Rosa to groom him. Rosa does not groom, but scratches instead. The juvenile male walks away and gives a neck-bite to a female B who happens to pass by. Rosa looks at Rossini, then approaches and grooms him.

One observation of Cadet indicated that Cadet also tended to prevent his initial females from interacting with immatures:

A female 2 of Cadet has walked away from her initial leader and joined a play group behind a little bush. Suddenly Cadet appears and chases after the female, who hides in the bush, screaming. Cadet tries in vain to grasp the female. Finally, the female emerges from the bush, still screaming. Cadet grabs her and bites her on the neck.

F. Discussion

The development of the initial units in band I of Cone Rock troop occurred as a gradual process. Two phases could be distinguished: short but regular interactions between young subadult males and female infants brought together prospective members of an initial unit (onset phase); in a later stage the male separated the female from her parental unit and the initial bond became an autonomous unit. Several triadic effects postulated in section 4F were found to be parts of the complex process of forming an initial unit.

The development of the initial bond appeared to be delayed because of the harem leader's influence. This effect led to a deviation from the dyadic rules given in chapter 4. The present chapter has provided qualitative evidence that harem leaders occasionally interfered with interactions between subadult males and immature females. Two interventions of Rosso led to a temporary breakdown of Rossini's initial bond. However, the main cause which accounted for the delayed forming of an initial unit seemed to be an inhibitive effect that the relationship between harem leaders and their daughters exerted on bachelors. In the experiments of Kummer et al. (1974), one casual interaction and the spatial proximity of two animals were sufficient to produce inhibited behavior even in rivals that had never seen the pair before. In the field the triad of a harem leader, a follower, and an immature female does not consist of strangers. A female infant who is born into a one-male unit, and who lives there with her mother, appears sufficiently identified with regard to her association with a harem leader. Spatial proximity and casual interactions with the harem leader and the harem leader's tendency to herd his daughters might then be sufficient to inhibit followers in their interactions with immature females.

A dyad consisting of a harem leader and an immature female appeared to inhibit a follower more strongly than a dyad consisting of a harem leader and an immature male. That harem leaders were not seen to herd immature males and that immature males stayed farther away from their parental units might explain the difference. On the other hand, immature females interact less frequently with and stay farther away from harem leaders than do adult females. Therefore, a harem leader and an immature female might be less inhibiting to followers than the harem bond. According these triadic effects bachelors can interact easily with male partners, less easily with immature females, and scarcely at all with adult females.

The possessive claim that harem leaders seem to have on immature females may be the cause of the follower's subtle but persistent techniques trying to interact with immature females. Otherwise the follower risks provoking the intervention of the harem leader.

When the initial unit reaches autonomy both the initial leader's and the female's social relationships change as a consequence of further triadic effects: The initial leader interacts less frequently with male partners, a change which already begins during the onset phase of forming an initial unit. Only notifying, a characteristic form of presenting among males, becomes more frequent in initial leaders. The female's relationship with her mother and her interactions with immatures are reduced in frequency. The establishment of the initial unit decreases the frequency of conflict behavior for the initial leader as well as for the initial female.

The gradual development of forming an initial unit, which was observed in band I of Cone Rock troop, is probably not typical for all hamadryas bands. Kummer (1968a), who first described the initial unit in the White Rock troop, noticed no gradual transition from the follower to the initial leader in his sample of eight identified units. On the contrary, it seemed that the initial leaders had acquired their females by means of a sudden action (Kummer 1968a and personal communication). Our own broad-sample data from Cone Rock and White Rock showed troop-specific differences that we have interpreted as different probabilities of forming initial units (section 3E). The close-study males of band III at Cone Rock showed no behavior that could be considered part of family forming. Two males acquired initial females in a sudden takeover. These observations agree with the description of Kummer. At present we do not understand such band-specific differences. A glance at the band compositions of Cone Rock troop (tab. 3) invites an apparently plausible hypothesis. Band III had a relatively low number of immature females, especially in proportion to the number of subadult males (13 immature females and 9 subadult males). One might at first hypothesize that a high number of immature females per bachelor favors a gradual development of the initial unit, whereas relatively few immature females promote an abrupt takeover. The hypothesis, however, does not agree with the observations of Kummer (1968a) because the composition of the larger White Rock band did not contain a particularly low number of immature females (16 immature females and 2 subadult males; tab. 3). Neither does it explain why Pirat, who was the follower of a large one-male unit with 5 immature females in band III, did not build up an initial unit gradually. The band-specific differences in forming initial units suggest that the bond that associates a follower with a one-male unit is not only or necessarily based on the follower's interest in juvenile females of that unit.

6   From the Initial to the Mature One-Male Unit

The initial unit is the nucleus of the future one-male unit, but it must yet be enlarged to reach the stage of full development. Kummer (1968a, p. 75) assumed that initial leaders enlarge their units by taking over females from old leaders. The social changes that took place in band I during the last months of our field study permitted us to learn more about the life cycle of family units.

A. The Acquisition of New Females

Between July 1971 and March 1972 we noted 5 instances of females who were taken over by new leaders in Cone Rock troop:

The first observation was made in band II. The young harem leader Marco had a one-male unit consisting of an old adult female with infant and a female 2. One evening Marco arrived at Cone Rock with 2 additional females—one adult female and one female 2. The new female 2, Susi, had been marked one month before. All we knew about her was that she belonged also to band II. Ten days before she had been observed in a one-male unit consisting, besides herself, of an adult leader, 2 adult females, a male 1, a male B, and an infant. At that time we had no idea about the composition of band II; we did not know, therefore, what had happened to the former leader of Susi. Marco had no wounds suggesting a fight over the females, and he did not notify any adult male who might have been the former possessor of the females. (How new possessors of females notify the former possessors will be described in chapter 7).

The second change occurred in band I and concerned the one-male unit of Bishop. Bishop was a young harem leader like Marco. His one-male unit comprised 2 adult females and the female infant Prima (tab. 19). One evening Bishop arrived at Cone Rock with 3 adult females. The new adult female had obviously been taken over from another band, or even from another troop, because all one-male units of band I were complete. Two days later Bishop had already lost the new female. The vigorous herding of Bishop had left naked spots on the head and the back

118

of the new female, so that she could have been identified easily. We did not find her again in any band of Cone Rock troop.

The third change was noted in band III. A one-male unit, previously with nine adult females, had only six left. We could not reconstruct the events.

The fourth change occurred in band II and again concerned the one-male unit of Marco:

19.4.1972: In the evening Marco arrives at the rock with a new female 1. Whenever an adult male passes near him he crouches over the female and bites her on the neck. Marco appears most excited when in the vicinity of Astrian, a harem leader with 6 adult females in the same band. Later Astrian notifies Marco. We presume that Marco has taken over the female 1 from Astrian, but we cannot verify it because we do not know the precise composition of Astrian's harem. The female 1 succeeds in running away from Marco and gaining a tiny ledge in a steep cliff where Marco cannot reach her. Marco threatens her, but after 25 minutes the female still has not left her safe place. Meantime it has become dark. Marco does not move to his usual sleeping ledge; instead he settles 2 meters from the female 1.

20.4.1972: In the morning the female 1 is not near Marco, but near Brutus. Brutus is a young adult male, the initial leader of 2 juvenile females and a follower of Astrian. Brutus is groomed by the female 1. Then the female returns to the one-male unit of Astrian. We cannot observe what she does there because the one-male unit of Astrian walks out of sight. After 10 minutes the female leaves the harem of Astrian. Brutus chases after her, and she runs back into Astrian's one-male unit. Brutus walks away; she follows him. He settles down and looks at her. She stops 5 meters before him, utters staccato coughing, and then slaps the ground while looking at Brutus. Brutus looks away. The female withdraws, then approaches and presents to Brutus, who mounts her. She grooms Brutus. Astrian does not respond to this; possibly he is cautious because 5 of his 6 adult females are in estrus. Marco sits with his one-male unit behind Cone Rock and does not approach Brutus.

In the evening, however, Marco has won back the female 1 from Brutus. He herds her aggressively. She now appears more ready to follow her new leader. Marco settles with his one-male unit on the usual sleeping ledge. The female 1 remained in the one-male unit of Marco.

The fifth incident occurred again in band I. One evening several males of band I climbed acacia trees, shook branches and uttered ohu-roars. In search of the cause we found that Spot had walked away from his band and had approached band IV, which was returning from the foraging march. An adult leader of band IV was just chasing after one of his females, biting her on the neck. Meantime Spot chased after another female of the same leader. The leader immediately lunged at Spot; a short fight took place, then the leader ran away with his females. Spot returned to his band alone.

The behavior of Spot and Bishop tempted us to speculate that initial and young harem leaders acquire new females mainly from other bands. The speculation seemed plausible because it suggested a compensation for the formation of initial units, which takes place within the band. But then things turned out differently:

> In the evening of 3.5.1972 the old leader Padorn sits with his complete one-male unit at his usual place. Padorn's mantle is blood-stained, one eye is swollen, and one arm wounded. About 10 meters away Spot, his oldest follower, lies sphinxlike on the ground, completely immobile. The other one-male units of the band are arranged in a semicircle around Padorn and Spot. Several times Padorn stands up, turns on the spot, and sits again. Four times the old leader notifies his follower Spot. One of his adult females moves with her leader and seems anxious to stay as close as possible to him. Suddenly Padorn starts pumping cheeks and yawning while looking repeatedly in the direction of Spot. Rosso threatens Spot in the same way. Admiral and Cadet yawn. Bishop notifies Spot. Elizabeth, the initial female of Spot, sits in 3 meters. She approaches Spot and presents to him. Spot touches and sniffs at her anal region. After 10 minutes of pumping cheeks, Padorn notifies and returns to his place.

We interpreted the situation as an unsuccessful attempt by Spot to take over an adult female from Padorn. Rosso and Admiral seemed to assist Padorn by threatening Spot. The interpretation was confirmed a few days later:

> Padorn sits alone. He has lost his two adult and two immature females. Scars in his face and a fresh wound on his hind leg bear witness of a heavy fight. One of Padorn's former adult females stays with Spot, whose face scars show that he had the decisive fight with Padorn. The other adult female of Padorn has been taken over by Cadet, who shows no signs of a fight. Freundin, the prospective initial female of Freund, was lost and never found again. Instead, Padorn's 2 immature females, his daughters, are now with Freund.

The breakup of Padorn's harem is graphically summarized in figure 21. The female whom Spot has taken over was the same individual who had seemed anxious to stay near Padorn after the first fight between Padorn and Spot. The dividing of Padorn's females has separated the 2 immature females from their mothers.

Three weeks later the next change took place: Rossini took over Elizabeth, the initial female of Spot. Spot seemed so absorbed in his newly acquired adult female that he did not pay sufficient attention to his initial female, and Rossini had taken advantage of this. Fresh wounds on Spot and Rossini indicated that a fight over the female had taken

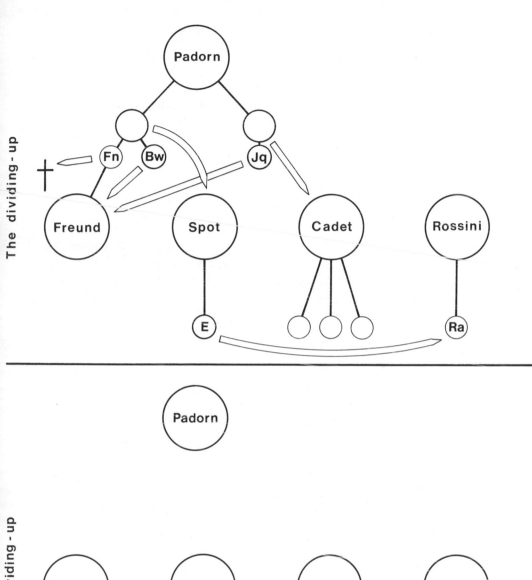

Figure 21. The breakup of Padorn's one-male unit in May 1972. Big circles represent mature males who were involved in the process; medium circles are adult and small circles immature females. Fn = Freundin; Bw = Browny; Jq = Jacqueline; E = Elizabeth; Ra = Rosa. Arrows indicate transfers of females; the cross represents the loss of a female during the process.

place. A few days later, it seemed that Rossini had even attempted to take over the adult female from Spot (probably Spot tried to reconquer Elizabeth and risked losing his adult female). The interaction pattern observed after the first fight between Padorn and Spot (see above) appeared again: Spot seemed very excited and intensively herded his adult female; several times he notified Rossini, and a fight between Spot and Rossini arose.

Soon after these spectacular events we left Ethiopia. When I visited Cone Rock in April 1973, I found that the series of changes had continued. The one-male units of Admiral and Rosso had also been dissolved. These major events, which left no one-male unit unchanged, had taken place during our absence. Figure 22 is a reconstruction of the events. As in the case of Padorn and Spot, the breakup of Admiral's and Rosso's one-male units had involved fights. The defeated leader Admiral had a long hole on the bridge of his nose. Rosso's face was scarred. The fact that his scars appeared fresher than those of Admiral indicates that the breakup of Admiral's one-male unit preceded that of Rosso's unit. The main recipients of Admiral's and Rosso's former females were Cadet and Bishop, who probably initiated the fights with the old leaders. Cadet and Bishop showed no signs of the fights. This is not amazing, because in the case of Padorn and Spot it was also the defeated leader who was more severely wounded, whereas Spot's scars had disappeared soon. During this second series of breakups the one-male unit of Bruno disappeared. We guess that Bruno became involved in fights over females and might have been severely wounded, because he had the handicap of a damaged eye.

It is striking that Cadet and Freund, who had acquired females from the former one-male unit of Padorn in 1972, lost these females to Spot by 1973. Thus, in the end, Spot had all the former females of his leader Padorn, who, in the meantime, had died.

Within one year the dominant harem leaders of band I, Padorn, Admiral, and Rosso, who had possessed 15 of the 19 adult females, lost nearly all their females. Only Rosso still had one left. The former initial leaders Spot and Cadet and the young harem leader Bishop were now the dominant harem leaders, owning 11 of 16 adult females. Padorn was defeated by his oldest follower, Spot. Admiral presumably was also defeated by his oldest follower, Cadet. The suspected initiator of the fight with Rosso was the young harem leader Bishop. We guess that Bishop was a former follower of Rosso. The rule would then be that the oldest followers deprive their harem leaders of the adult females. We shall return to this aspect in chapter 7.

A common feature in the enlargement of family units was that the new females were acquired during the foraging march. At the rock we registered only the results of the events but neither the fights nor the takeov-

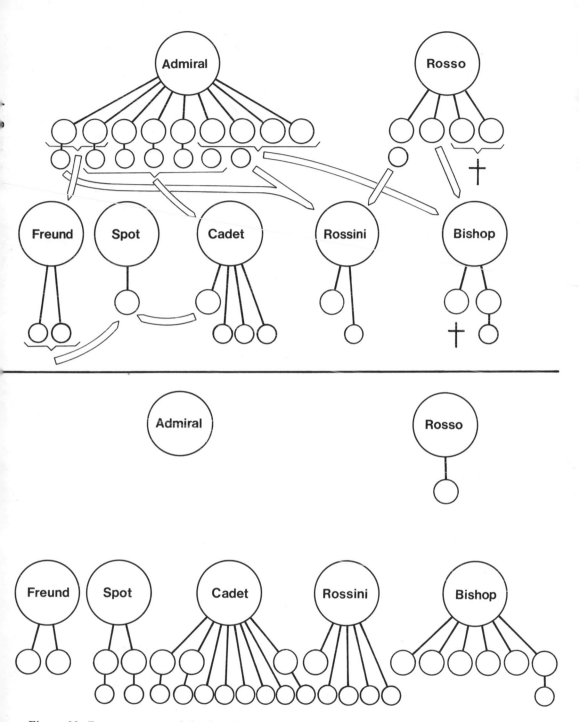

Figure 22. Reconstruction of the breakup of Admiral's and Rosso's one-male units, which took place between August 1972 and January 1973. Big circles are mature males who were involved in the process; medium circles are adult, and small circles imma-ture females. Lines between adult and immature females mark mother-daughter rela-tionships; lines are missing where the mother could not be identified. Arrows indicate transfers of females and crosses the loss of females.

Table 25. Redistribution of Females in Band I between August 1971 and February 1974

Status of recipients of females	Recipient's relationship to the loser of female:			Distribution within the band according to status of recipient
	Follower of loser	Member of same band	Member of other band	
Old harem leaders (Padorn, Admiral, Rosso)	0	0		0
Oldest followers (Spot, Cadet, Bishop)	15	4		19
			8	
Intermediate followers (Freund, Stupsie, Rossini)	2	5		7
Youngest followers (Coci, Minimus, Pepsi)	0	0		0

The redistribution includes transfers of females which took place through forming initial units or breaking up one-male units. The null hypothesis is made under the assumptions that each mature male of Cone Rock troop had equal chances to receive one of the females, that losers could win back females, and that the females who disappeared were taken over by males of other bands. The null hypothesis is rejected in the following respects: Main recipients are the oldest followers ($X^2 = 38.08$, $p < 0.001$); females tend to remain within the band ($X^2 = 42.86$, $p < 0.001$); within the band they tend to be taken over by followers of the losers ($X^2 = 10.88$, $p < 0.001$).

ers. Possibly the attention of the harem leaders during the day is directed to the coordination of travel and therefore diverted from their females. Also, distances between individuals become greater on the foraging march, and the environment contains more bushes and trees, which limit the range of visibility. On 3 of 15 foraging marches, during which we followed band I for one hour each time, fights occurred; but in all 3 cases the limited visibility prevented our recognizing their context.

Table 25 summarizes the whole redistribution of females that took place between August 1971 and February 1974 in band I. The table shows that females tended to be transferred within leader-follower associations and that the oldest followers were the main recipients of females. In summer 1975 the 2 juvenile females of Spot went back to Freund, who was their original recipient after the breaking up of Padorn's one-male unit. If we include this transfer in the entire reshuffling process we may conclude that oldest followers tended to receive adult females and the second followers juvenile females ($X^2 = 4.55$, $p < 0.05$).

B. The Enlarged One-Male Unit

1. The Weakening of the Initial Bond

Some changes which are correlated with the transition from the initial to the mature one-male unit can be analyzed for the units of Spot, Cadet, and Rossini. The unit of Freund is not taken into consideration, because the enlargement of his unit coincided with the loss of his initial female, so that changes of the initial bond cannot be studied. The unit of Bishop also is excluded; Bishop did not belong to the close study sample, he already was a harem leader before the enlargement of his unit and the enlargement took place after our main study period.

One common feature of the enlargement of Spot's, Cadet's, and Rossini's initial units was the conspicuous weakening of the initial bond. Frequencies of grooming between leaders and females before and after the enlargement are summarized in table 26. The table shows the following changes which took place with the enlargement: (1) In all 3 units grooming between leader and initial female decreased in frequency after the enlargement of the unit (Spot: $X^2 = 62.00$, $p < 0.001$; Cadet: $X^2 = 24.16$, $p < 0.001$; Rossini: $X^2 = 66.24$, $p < 0.001$). In particular, none of the 3 leaders groomed his initial female after the enlargement. (2) All 3 leaders more frequently groomed their new female than they had groomed their initial female before the enlargement (Spot: $X^2 = 18.73$, p. < 0.001; Cadet: $X^2 = 41.42$, $p < 0.001$; Rossini: $X^2 = 21.58$, $p < 0.001$). This result agrees with the statement by Kummer (1968a) and Kummer et al. (1974) that leaders groom their females more frequently when the male-female bond is new or threatened. (3) None of the new females groomed their leader more frequently than the initial females had before the enlargement. Therefore, the weakening of the initial bond is caused by the leader, who is absorbed in interacting with the newly acquired female and neglects his initial female.

A second feature was common to the 3 enlarged units. The initial females, who were released from their leader's constant attention, attracted bachelors who attempted to become followers of the enlarged one-male units:

Table 26. Grooming within One-Male Units before and after Their Enlargement

it	Grooming in the initial unit		Grooming in the enlarged one-male unit			
A: R:	I-leader I-female	I-female I-leader	H-leader I-female	I-female H-leader	H-leader New female	New female H-leader
ot	3.0	24.1	0	0	7.8	18.3
det	0.3	39.8	0	22.4	4.7	2.2
ssini	6.7	17.1	0	8.1	14.7	23.1

e table shows that grooming between leader and initial female decreased after the largement of the unit and that leaders frequently groomed their new female (for test lues see text). A = actor; R = recipient; I = initial; H = harem.

In the case of Spot, it was Stupsie who was attracted by the weakened initial bond of Spot with Elizabeth. So far, Stupsie had been a follower of Admiral. Before the enlargement of Spot's initial unit Stupsie was found twice in 9 months to stay in the vicinity of Spot's initial unit, and no interaction between him and the initial unit was observed. After Spot had taken over the adult female, Stupsie was found 8 times in 3 weeks in the vicinity of Spot. On the first morning we recorded Spot as having acquired a new female, Stupsie sat already near the enlarged one-male unit:

> Stupsie sits 1.5 meters behind Spot. Then he returns to the one-male unit of Admiral. When the band moves on the rock, he again joins the unit of Spot. Spot, who has also moved, sits down. Stupsie stops behind him, yawns, then presents to Spot and sits 2 meters away.

In the subsequent days Stupsie repeatedly attempted to approach Elizabeth, the initial female of Spot, and to interact with her. Once Elizabeth showed an intention to present to Stupsie but then rushed to her leader and groomed him. On another occasion Elizabeth walked away from Spot. Stupsie approached her and presented her his flank to be groomed. Elizabeth groomed him for a few seconds. While Spot showed less grooming with Elizabeth, he now started to herd her—the first time we saw him do so. Stupsie did not remain a follower of Spot, but returned to the one-male unit of Admiral. A few days later Elizabeth was taken over by Rossini.

The enlarged unit of Cadet was joined by the young subadult Minimus, who so far was a follower of Admiral. He was found in 3 of 19 records to stay near Cadet before the enlargement of Cadet's unit, and in 8 of 11 records after the enlargement. The first day after the breakup of Padorn's one-male unit Cadet's initial females did not stay with their leader, but were found in the vicinity of Minimus. Before Cadet had taken over the new adult female, Minimus had been seen to occasionally groom with the initial females of Cadet; now these interactions seemed to have become more frequent (since neither Minimus nor Cadet's initial females belonged to the close-study sample, the difference could not be measured). Until the end of the main study, Minimus followed both the one-male units of Admiral and of Cadet. During my visit in 1973, when the one-male unit of Admiral had been dissolved, Minimus had definitively become a follower of Cadet. He neither interacted nor moved with the defeated leader Admiral as did Coci (see section 6G). In 1974 his relationship with one of Cadet's initial females was approaching the stage of an autonomous initial unit.

In the case of Rossini it was Pepsi who was attracted by the enlarged unit. Before the enlargement of Rossini's initial unit Pepsi was never observed to follow the initial unit. Once he had been observed to approach and mount Rosa, the initial female of Rossini, but the interaction had occurred while Rossini was prevented by an intervention of Rosso to interact with Rosa. Interactions between Rossini and Pepsi occurred only during the period in which both males were interested in the female Blondy (section 5D). On the very evening of Rossini's acquisition of Elizabeth Pepsi already followed Rossini's initial female Rosa throughout the evening. The next day he was also found near Rosa. In the following morning Pepsi presented to Rossini. Then he approached Rosa who had walked away from her leader and prevented her return to Rossini by interposition. When the band left for the foraging march, Pepsi again presented to Rossini. In the evening Pepsi climbed the rock with the one-male unit of Rossini. At the top of the rock he attempted to lead Rosa: As soon as Rosa walked away from Rossini, Pepsi took the place in front of her and "led" the female while looking back at her. In the following 2 days Pepsi remained in the vicinity of Rosa, and twice he presented to Rossini. Then he returned to his leader, Bishop, where he remained for the remaining 2 months of our study.

From the 3 bachelors who attempted to become followers and to interact with the neglected initial females, only Minimus finally succeeded in building up his own initial unit in this way. The results of chapter 7 will permit us to hypothesize on why exactly these bachelors associated with the enlarged one-male units and why only Minimus succeeded.

2. The Behavior of the Transferred Females

Since most of the transferred females did not belong to the close-study sample, and since most transfers occurred after the main study period, the behavior of the transferred females was not investigated systematically. Qualitative impression, however, suggest that transferred females attempted to approach animals from their former one-male units, among them their former leader and their children, and that transferred females tended to establish grooming relationships with unfamiliar females in the new one-male units rather slowly:

The constant attention of the new leaders did not leave much freedom of movement to the newly acquired females.

A few observations, however, indicated that some transferred females attempted to approach their former leader. One day after the breakup of Padorn's one-male unit the new adult female of Spot attempted twice to withdraw from Spot and to approach her former leader, Padorn. Spot immediately interposed and once led his new female near Padorn, where he mounted her. In the next evening a similar sequence occurred. Whether the second former female of Padorn, who was taken over by

Cadet, behaved in the same way or not, escaped our attention. The only observation was made 3 days after the transfer on the foraging march: Cadet's new female shortly veered off in Padorn's direction when the latter deviated from the travelling direction of Cadet's unit.

In 1973 we made the following observation on Freund's adult females, who were presumably taken over from Admiral:

> Freund sits with his two adult females on the rock. Admiral sits 2 meters away. One of Freund's females withdraws slowly from Freund, cautiously approaches Admiral and sits 0.3 meters from Admiral, who does not respond. The second female also walks away from Freund and sits close to Admiral. Freund does not react. After a few minutes he passes Admiral and sits down. His females return to him. Freund gives a neck-bite to one of them.

The following observation suggests that one of Freund's adult females attempted to approach females who also had belonged to the one-male unit of Admiral:

> Freund sits with his two adult females on the rock. The one-male unit of Cadet, which contains former females of Admiral, passes Freund. One of Freund's adult females follows Cadet's unit, looking repeatedly back at Freund, who does not react. She settles 10 meters away from Freund and 6 meters from Cadet. A male 1 from Cadet's one-male unit approaches her and is groomed by her. Freund approaches. The adult female withdraws from him while the male 1 presents to Freund. Freund ignores the juvenile male and sits down. The adult female has approached Cadet, sits 4 meters from Cadet and looks at Freund. After a few minutes Freund approaches her, and she presents to him. Freund sniffs her anal region. The female again approaches Cadet while looking at Freund. Cadet also looks at Freund and notifies him. The one-male units of Cadet and Freund remain settled side by side.

In the enlarged one-male unit of Bishop I observed a few times that one or two of the new adult females were 10 to 12 meters away from Bishop and interacted with immature members of their former one-male units. Possibly the immatures were children of the adult females, but we could not verify the assumption because we could not identify individually the animals involved.

We could, with more or less certainty, identify 4 mother-daughter pairs which were separated through the breakup of Padorn's, Admiral's and Rosso's one-male units, both the mother and the daughter remaining in band I: After the breakup of Padorn's one-male unit the two adult females were taken over by Spot and Cadet, respectively, while their daughters, Jacqueline and Browny, were taken over by Freund (fig. 21). At that time Jacqueline was a female 1 and already weaned; Browny was still in age class B and not yet weaned. Freund, their new leader, showed an ambivalent behavior toward them. On one hand, he herded them aggressively, as the scars on the females' heads demonstrated (see also Kummer 1968a, p. 70). On the other hand, he appeared peculiarly apathetic and showed no reaction when the females walked away from him on the rock. Browny took advantage of this freedom of movement and attempted to return to her mother, who was now Spot's female. One week after the separation from her mother Browny approached her as far as 0.3 meters, but did not dare to move closer. One month after the separation we observed her as she was groomed by her mother. The following observation indicates that Browny's separation from her mother was probably not maintained only by Freund's herding behavior, but also by Spot's excluding behavior:

Browny's mother sits 1 meter from Spot. Browny has approached hesitantly, sits close to her mother and grasps one of her nipples. Spot stares at Browny and looks at her mother, who approaches him and sits 0.5 meters from him. Browny follows her, sits near her and again grasps a nipple. Spot yawns. After one minute the mother moves away from Spot, and Browny follows her. Spot looks at the mother and walks away. She presents to him and follows. Browny remains seated 2 meters away and then returns to her leader Freund.

After 11 minutes Browny again approaches her mother, who has stopped grooming. Spot scratches. Browny sits close to her mother and is groomed by her. Spot scratches again. After 5 minutes' grooming Browny finally sucks. Spot does not respond any more to the mother-daughter interaction.

This is the first observation of a leader tending to exclude a female from his one-male unit!

Jacqueline showed no attempts to approach her mother, who was taken over by Cadet. At the time of my field visit in 1973 Spot had taken over Jacqueline's mother from Cadet, and Freund had acquired two adult females (fig. 22). Both Jacqueline and Browny were again united with their mothers in the one-male unit of Spot.

In the third case of mother-daughter separation we are less certain about the individuals' identity, because neither mother nor daughter was marked. During my short field visit in 1973 I observed 3 times that one of Freund's adult females walked away from her leader and groomed a young female 1 of Cadet's one-male unit. By 1974 this adult female of Freund was associated permanently with a female 1. We presume that the juvenile female was the daughter of the adult female and that she succeeded in joining her mother as did Browny and Jacqueline.

The fourth case involved the second adult female of Freund. Her daughter seemed to have been taken over by Rossini. The mother-daughter relationship is inferred from the following observations:

27.4.1973: In the evening Freund sits with one adult female on Cone Rock. The older of his adult females sits a few meters from Rossini's one-male unit, which has settled on the small sleeping cliff opposite Cone Rock, 100 meters away from Freund. We do not see the adult female interact with any of Rossini's animals.

28.4.1973: In the morning the old female of Freund is still on the small sleeping cliff opposite Cone Rock. She seems to have passed the night there with the one-male units of Rosso and Rossini. Before band I leaves for the foraging march, Freund climbs with the remaining adult female over to the small cliff and returns with both adult females.

In the evening Freund arrives at Cone Rock with only one adult female. The one-male units of Rosso and Rossini have again climbed the small sleeping cliff and settled there. In their vicinity sits the old adult female of Freund. A female 1, whom Rossini has taken over from Admiral, approaches her. The old female invites the female 1 to groom her. The female 1 looks at Rossini and then grooms the old female. Rossini approaches and the female 1 immediately stops grooming. Until dark the old female was found to remain on the small sleeping cliff.

During the remaining 3 weeks of the field visit the old female of Freund was never seen again to approach the one-male unit of Rossini. In 1974 she was still in the unit of Freund, and the female 1 who had groomed her was still in the unit of Rossini.

The above-mentioned observations suggest a cohesiveness among females who have lived in the same one-male unit. The following observations illustrate that this cohesiveness was accompanied by a reluctance to establish grooming relationships with unfamiliar females in the new one-male unit.

The enlarged one-male units of Rossini and Bishop brought together females who had been living in different one-male units before. In these cases the females had to clarify their rank positions within the new unit. Before Rossini acquired Elizabeth from Spot, he was occasionally groomed by the female 1 from Rosso's one-male unit. Elizabeth, the newcomer in Rossini's unit, had to establish her rank position with reference to Rosa and this female 1:

28.5.1972: Elizabeth has been taken over by Rossini.

7.6.1972: The one-male units of Rosso and Rossini have settled side by side on the rock. The unit of Rosso moves, but the female 1 from Rosso's unit remains sitting in front of Rossini and is groomed by Elizabeth. Rosa stands 0.8 meters away from the grooming pair and threatens the female 1. The female 1 and Elizabeth approach Rossini and groom him. Rosa walks away.

26.6.1972: Elizabeth grooms Rossini. The female 1 from Rosso's unit approaches Rossini and attempts a few times to groom him, but Elizabeth chases her away. Rossini does not respond.

29.6.1972: Elizabeth sits close to Rossini. Rosa approaches them, followed by the female 1, and sits in front of Rossini. Rossini invites Rosa to groom him and Rosa grooms. The female 1 intends to groom Rossini as well, but Elizabeth chases her away. The female 1 runs around Rossini and succeeds to groom him. Elizabeth attempts to chase her away, but the female 1 remains seated. Elizabeth then also grooms Rossini.

10.7.1972: Elizabeth grooms Rossini. Rosa approaches and intends to groom her leader. Elizabeth threatens Rosa. Rosa screams and presents to Rossini, who does not respond. Elizabeth continues to threaten Rosa. Both females run around Rossini. Rosa is still screaming. Finally, Rossini bites Rosa on the neck. She sits down and grooms him.

12.7.1972: Rossini moves with Elizabeth and Rosa. When Rossini and Elizabeth sit down, Rosa approaches them, shows an intention to present to Elizabeth, then sits close to Rossini.

In the 2 months during which we could observe the enlarged one-male unit of Rossini no grooming interactions were noted between Elizabeth and Rosa, while Elizabeth groomed at least once with the female 1 from Rosso's unit. This observation agrees with one of the rules which Kummer (1975) had formulated for captive geladas: The less compatible a pair type is in isolation, the greater is its tendency to regress in the presence of others. The smaller age difference between Elizabeth and Rosa suggests a lower a priori compatibility than does the greater age difference between Elizabeth and the female 1 from Rosso's unit.

The enlargement of Bishop's one-male unit also brought together females who had been living in different one-male units before (fig. 22). During the field visit in 1973 I repeatedly observed that Bishop's females were quarreling and fighting among each other. Whether grooming occurred between females who came from different units could not be decided.

The enlargement of Marco's one-male unit in band II (see section 6A) led to grooming interactions which also agree with Kummer's above-mentioned rule: the adult female, Martha, the female 2, Biancha, and a female infant belonged to the unit of Marco from the beginning of our study. The adult female, Sana, and the female 2, Susi, were later acquired together by Marco. During the 8 months since the enlargement of Marco's unit no grooming interactions were observed between the females 2 Biancha and Susi. Neither do our records contain any grooming between the two adult females. At the same time grooming between the adult females and the females 2 reflected the persisting affinity among females who had been living within the same one-male unit: Biancha groomed 5 times more frequently with Martha than with the new-comer Sana, and Susi groomed 23 times more often with Sana than with Martha.

After the enlargement of Cadet's unit one former adult female of Padorn was brought together with Cadet's initial females, who came from Admiral's unit. The integration of the adult female went smoothly and three days after the transfer the adult female was already grooming with one of Cadet's initial females. This again agrees with Kummer's above-mentioned rule.

Finally, in the case of Spot, the new adult female came from the same one-male unit than Spot's initial female, Elizabeth. Two days after her takeover, the adult female already groomed with Elizabeth.

3. Family Forming and Sexual Swellings in Females

Kummer (1968a, pp. 177–79) has pointed out that the estrus cycle of hamadryas females is affected by social factors. Females tend to synchronize their cycles within the one-male unit; Kummer (1968b) suggested that juvenile hamadryas females seem to have their first sexual swelling at an earlier age than do cynocephalus females and that the difference might be correlated with the earlier consort role of hamadryas females. While our age estimates do not support the assumption that hamadryas females start to cycle earlier than cynocephalus females (see Section D), our limited observations agree with the presumed effect of the consort role on the onset of swellings in hamadryas females.

During the main study period we identified in Cone Rock troop 3 adult and 5 juvenile females who were taken over by new leaders. The first adult female came into estrus a few days after her takeover. Since we had no history of her cycling, we cannot determine whether her estrus was affected by the change of one-male unit. The second adult female came into estrus a few days before her takeover. She then had a permanent swelling for 2 months. The third adult female, who appeared to be old, developed no swelling after the takeover. Three of the 5 juvenile females were in age classes B and 1 and therefore were too young to develop swellings. The fourth female, who was in age class 2, had a swelling two weeks after her takeover. Most probably it was her first one; one month before her takeover we had trapped and marked her, and 11 days before the takeover we had recorded her behaviour; both times she showed no signs of a swelling. The fifth female, Elizabeth, was taken over at the end of her first swelling.

The 8 females were transferred to 5 different males; 4 of these males already had females, namely 1 adult and 6 juvenile females altogether. It seems that some of the resident females also were affected by the enlargement of their one-male units: The adult female had a black infant and, therefore, did not cycle. Two of the 6 juvenile females, Elizabeth

and Rosa, had their first swelling after their unit was enlarged: Elizabeth
had her first swelling 9 days after her leader Spot had taken over an adult
female; Rosa developed her first swelling 6 weeks after her leader Rossini
had taken over Elizabeth and at the same time of Elizabeth's third swelling. Rosa and Elizabeth were the only females of their leaders before the
unit's enlargement. In contrast, the other 2 males who acquired new
females, Marco and Cadet, had more than one female before the enlargement of their units. This led to a conspicuous difference: The females of
Marco and Cadet stayed quietly aside and groomed within their units,
while their leaders vigorously drilled the new females. Elizabeth and
Rosa had no other females to groom with and, therefore, they appeared
to be more affected by the herding behavior of their leaders.

These limited observations suggest that herding behavior of males
may affect the development or duration of swellings not only in
transferred females themselves, but also in resident females.

Among the 8 juvenile females of the close-study sample Biancha was
the only animal whose first swelling occurred apparently without a con-
current change in the composition of her unit. This permits analysis of
some behavioral changes correlated with the occurrence of the first sexual swelling alone. Figure 23 shows some behavioral frequencies before
and after Biancha's first swelling. The graph demonstrates that after her
first swelling Biancha interacted less frequently outside her one-male
unit, but more frequently within the unit. One week after the occurrence of the swelling Biancha's head showed scars, which probably resulted from a neck-bite of her leader, Marco. Subsequently Marco was
occasionally observed to herd Biancha when she attempted to leave the
one-male unit. Before, we had never observed Marco herding Biancha.
With the first estrus, we may assume, a female becomes more strongly
integrated into a one-male unit. If this is true, the separation of an initial
female from her parental unit should take place before the occurrence of
her first swelling if the rivalry between the harem leader and the follower
over the female is to be kept minimal. Indeed, in all 6 initial units of
band I which had already passed or were in the phase of separation, all 8
initial females had not yet shown sexual swellings.

C. The Defeated Leader

The fate of the 3 defeated leaders, Padorn, Admiral and Rosso, had
some common traits. To begin with, their physical appearance changed
conspicuously; in addition to the scars, the wounds, and the transitory
swelling of the superciliary region resulting directly from the fight, their
faces turned darker, they lost hair and became skinny.

As harem leader, Padorn had a brown face. After the fight with Spot
and the loss of his females Padorn's face turned ash-gray, 3 weeks later it

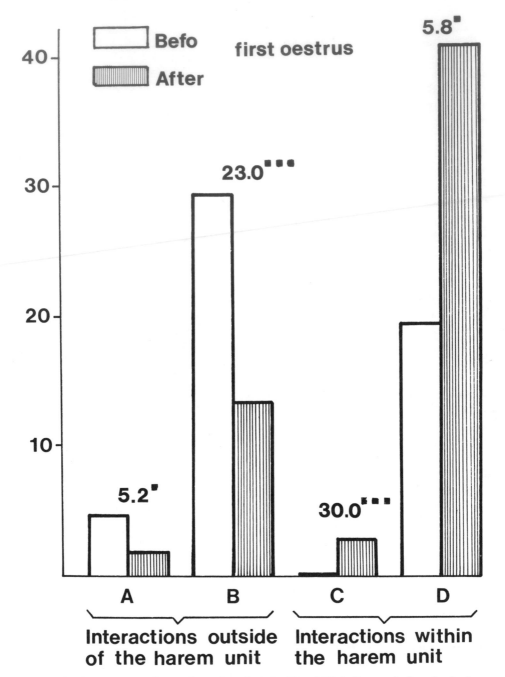

Figure 23. Interactions of Biancha, a female 2–3 of band II, before and after she had her first sexual swelling. The behavioral samples cover an observation time of 136 minutes before, and 101 minutes after the first estrus. A = play with immature males; B = grooming with immature males; C = grooming with female infant; D = grooming with adult female. Figures above the bars are X^2 testing the difference of frequencies; ▪ indicates a $p < 0.05$, ▪▪▪ a $p < 0.001$.

changed to dark brown. Admiral's face, which was violet-reddish before, turned a dark violet-brown after the loss of his females. The defeated leaders also lost part of their mantle hair. Admiral had nearly bare areas on his shoulder (fig. 24). From the rear the old males looked like subadults, only their heads appeared disproportionately large. Padorn and Admiral apparently lost weight which was manifested by loose abdominal skin folds. In 1973 Padorn had disappeared and probably died. Admiral survived the loss of his females and was still alive in 1974. By then he had regained the full thickness of his mantle; only the dark pigmentation of his face had remained unchanged.

Rosso, who had kept one adult female, changed his appearance in the same way as Padorn and Admiral, but the changes were less conspicuous.

The further social lives of the defeated leaders also had similarities:

(1) The loss of the females was an irreversible event. None of the defeated leaders ever attempted to win back or even approach his former adult females, nor did they acquire new females. For example, when the 2 adult females whom Freund had taken over from Admiral sat close to their former leader, Admiral remained completely passive.

(2) The oldest followers, who were the main recipients of females (tab. 25), tended to become independent of their defeated leaders. Spot, who had been Padorn's oldest follower, moved completely on his own after having defeated Padorn, but he kept his sleeping ledge near Padorn's. Padorn seemed to avoid the vicinity of Spot. In the first days after the loss of his females Padorn did not immediately occupy his usual sleeping place, but settled a few meters away at the edge of the sleeping area of band II. Cadet became independent of Admiral in the same way. For Bishop no such changes were noted because Bishop already was independent of Rosso before acquiring additional females.

(3) The second followers tended to remain in the vicinity of their defeated leaders. Padorn's second follower was Freund. On the first morning after the loss of Padorn's females Freund sat 1 meter away from Padorn. When band I moved on the rock Freund followed Padorn closely. For the foraging march the band traveled toward the north. Padorn did not follow the band, but took the direction to northeast. Only Freund followed him. During the next days Freund remained in the vicinity of Padorn; he changed his sleeping place, which was situated 3 meters below Padorn's, and slept close to Padorn. The 2 males, however, did not groom each other, probably because of the 2 juvenile females whom Freund had taken over.

The second follower of Admiral was Coci. In 1973, when Admiral had lost his females, Coci was still regularly found to move with and to sit near Admiral. Coci had acquired none of Admiral's females. Several

Figure 24 a and b. The defeated leader Admiral of band I. The figure illustrates the loss of mantle hair and the skinny appearance in defeated leaders. The hole on the nose results from the fight over the females.

times the 2 males groomed one another. Grooming between Admiral and Coci had not occurred during the time of Admiral's leadership.

If it is true that Bishop was a former follower of Rosso (section 6A), then Rossini was the follower second in age. Rossini, who had not taken over females from Rosso, regularly remained in the vicinity of his leader, and often the animals around Rosso and Rossini appeared as a single group.

(4) The defeated leaders interacted with their male and female offspring to the extent that such interactions were possible. The male and female offspring of the defeated leaders were rarely seen to interact with their fathers as long as adult females were present in the one-male unit. After the breakup of the unit, interactions between father and children became more frequent: After having lost his females, Padorn regularly groomed with his son Hajö, a male 2. It was the first time that we observed grooming between Padorn and Hajö. The following observations illustrate the apparent mutual interest of both father and son in these interactions:

Hajö grooms Padorn. Minimus, the subadult follower of Admiral, passes by. Hajö leaves Padorn, follows Minimus and sits 2 meters behind the subadult. Freund, who has been seated 2 meters away from Padorn, passes Padorn and settles 1.5 meters above him. Hajö runs back to Padorn, threatens Freund and starts grooming Padorn.

Band I sits in the dry riverbed in front of Cone Rock. Padorn lies in the sand. Suddenly some screaming is heard from a play group in about 20 meters from Padorn. Padorn immediately gets up and approaches the play group. There, Hajö stands and screams. Padorn approaches him, looks around and invites Hajö to groom him. Hajö starts grooming. A few minutes later Hajö joins a play group. Padorn again approaches him and invites his son to groom him. Hajö grooms again.

Padorn sits in the vicinity of Admiral's one-male unit. Hajö sits in front of him. The one-male unit of Rosso approaches and settles a few meters away. The male infant of Rosso's unit approaches Padorn. Padorn looks at the infant and smacks. The infant withdraws and approaches Rosso, who also smacks. Then the infant approaches Hajö, who embraces him. Rosso and Padorn start to pump cheeks and to yawn. Rosso lunges at Hajö and chases him away. Padorn runs after Rosso. There is a short fight between Rosso and Padorn, then both males settle 2 meters away from one another and threaten each other. Gradually they calm down. Padorn settles in the vicinity of Admiral's unit and starts self-grooming.

No interactions were observed between Padorn and his daughters, Elizabeth, Jacqueline, and Browny. As new females of Rossini and

Freund, respectively, Padorn's daughters probably had not enough freedom of movement to approach their father.

Admiral also groomed with some of his children after having lost his adult females. Several times he groomed with the juvenile males and females who belonged now to the one-male unit of Cadet. Probably the interactions between Admiral and his juvenile daughters from Cadet's unit were possible because Cadet's attention was absorbed by the 3 adult females whom he had taken over from Admiral. No interactions were recorded between Admiral and his 2 juvenile daughters whom Rossini had taken over. During the time of his leadership grooming between Admiral and the immature members of his unit had never been observed.

Even Rosso, who after all had kept one adult female, interacted with his male and female children. The 4 juvenile males and the one juvenile female stayed closer to their leader than ever before and even groomed him. It was impossible to decide whether the juveniles belonged to the unit of Rosso or of Rossini.

In other bands we found old males who resembled the defeated leaders of band I in physical appearance and social behaviour. In July 1971 we noted an old male without females in band II. He was severely wounded, could hardly move, and followed his band with difficulty. One morning he and a one-male unit apparently missed the departure of their band. The harem leader hurried to join his band and left the old male alone. This old male occasionally behaved in a childish way: he screamed with bare teeth and squealed like an immature animal. After a few weeks he disappeared and possibly died.

In June 1972 an old male without females joined band I. He had scars on his face, his superciliary region was swollen, his face was dark violet-brown and his mantle thin. The old male had been seen in band IV a few days before. He was found alternatively in bands I, III, and IV. He disappeared after a few weeks.

Also in June 1972 we found an old male without females in a band which occasionally visited Cone Rock and which contained an anubis female (see section 2G). The male also had scars, a swollen superciliary region, a dark face, and a thin mantle.

In January 1974 band I was joined by a solitary male from band III. This male did not appear particularly old. His face was dark and his mantle thin. Occasionally he was seen to scream and to squeal while embracing Stupsie, a subadult male of band I.

In bands II and III some of the leaders who were dominant males in 1971/72 had disappeared and former followers had now become leaders. In band II two of the 5 dominant leaders were not affected by the turnover and still had their females in January 1974.

Finally, there was an old male in band III, Zebra, whom we first saw in October 1971 and who was still alive at the beginning of 1974. Throughout this period Zebra was without females. He was never seen to interact with any females. He groomed with male partners in 22.8% of his observation intervals. This frequency was typical for bachelors, since the mean frequency of grooming with male partners was 25.1% for the close-study bachelors of Cone Rock troop (tab. 23). Among Zebra's grooming partners was Schoggi, a marked male 2 of band III. Intuitively we considered Schoggi all the time as a son of Zebra. Our assumption remained without support until the defeated leaders Padorn, Admiral, and Rosso were observed grooming with their sons. Zebra's frequency of conflict behavior (self-grooming, yawning, scratching, and wiping muzzle) was 8.7% of intervals. This was less than the mean frequency for the bachelors, which was 15.9%; but it accords with the mean frequency of conflict behavior in the initial leaders of band I, which was 8.4% of intervals (tab. 22). This result agrees with my interpretation that the high frequency of conflict behavior in bachelors was mainly caused by the bachelors' impeded access to females, whereas Zebra's behavior suggested that the old male was not interested any more in acquiring females.

Thus we found old males who showed the characteristic features of a defeated leader in all 5 bands which visited Cone Rock. This indicates that the defeat of old harem leaders might be the typical process of hamadryas bands for forming new one-male units.

D. Discussion

While the forming of initial units indicated band-specific differences (section 5A), the acquisition of adult females seems to take place in the same way in all bands. In the apparently typical process the oldest follower of a harem leader defeats the leader, who loses all or most of his females. The follower who wins the fight does not simply inherit the whole harem of his leader. Our observations suggest that the follower receives as many females as he can assert against his rivals. The rivals consist mainly of other followers of the band, but not of dominant harem leaders. After the decisive fight between Padorn and Spot, the remaining dominant leaders, Admiral and Rosso, made no attempt to acquire one of Padorn's females; Spot's rivals were the followers Cadet, Rossini and Freund. This agrees with the observation that competition for females occurred among followers before the breakup of Padorn's unit (see section 5D), whereas no competitive interactions were ever observed between harem leaders. This difference will be further supported in the next chapter.

The enlargement of initial units opened new possibilities for younger

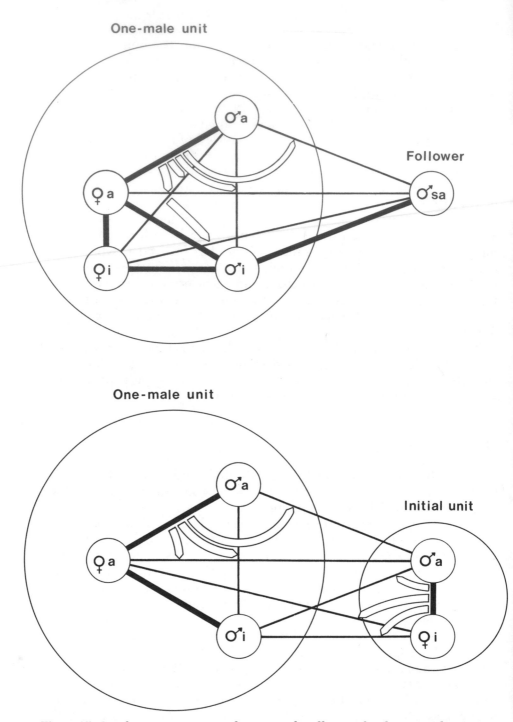

Figure 25. Graphic representation of some triadic effects in family units. The upper graph shows the sex-age classes of a one-male unit with a follower, the lower graph of a one-male and an initial unit. i = immature; sa = subadult; a = adult. The graph presents dyadic relationships which appeared not affected by triadic effects (————) and relationships which were found to be affected by triadic effects (————). The presumed causes for the suppression of interactions are marked with arrows.

bachelors to acquire initial females: The acquisition of adult females weakened the initial bonds of the oldest followers and, as a consequence, initial females became more easily accessible to bachelors. The bachelor Minimus formed his initial unit by acquiring one of Cadet's neglected initial females, and Rossini acquired his second female, Elizabeth, in the same way. When a one-male unit is broken up the initial leaders probably concentrate on taking over adult females, and this might permit younger followers to take over juvenile females directly from the dissolved one-male unit, as did Freund in band I.

With the investigation of family forming, we have attained the level of triadic rules according to the bonding-stage model, and, herewith, some access to the casual dimension of social structure. On the basis of our data of chapters 4 to 6 we postulate that 3 types of male-female bonds have an inhibitive effect on a variety of other social relationships (fig. 25):

a. The *harem bond* between adult male and adult female appears to have three inhibitive effects. Firstly, it suppresses grooming interactions of other mature males, among them of followers, with the harem leader and the adult females of the one-male unit. This effect has been tested experimentally by Kummer et al. (1974). It agrees with the following of our observations: followers groom with defeated leaders, if and only if they did not acquire or already have females; and, of course, they groom with the former females of the leader, as soon as they have acquired them. Secondly, the harem bond suppresses grooming interactions between harem females and animals outside the one-male unit (see Kummer 1968a, p. 80). Thirdly, within the one-male unit the harem bond suppresses grooming interactions between the leader and the immature members of the unit (see also Kummer 1968a, p. 81). Such interactions emerge after the leader has lost his adult females.

b. The *initial bond* in the autonomous initial unit has similar inhibitive effects as the harem bond: It reduces the initial leader's interactions with other males and the initial female's interactions with animals outside the initial unit. These effects were suggested by the troop differences in the broad sample (section 3E), by the deviations from the dyadic rules in the close study (section 4D), by the comparison of close study bachelors and initial leaders (tab. 23) and by the comparison of initial and noninitial females (tab. 24).

c. The *father-daughter bond*. Although interactions between harem leaders and their daughters were rare, occasional interactions and the possessive claim of leaders on immature females might be sufficient to make the father-daughter bond itself a suppressing factor for interactions between followers and immature females. This effect was suggested by the deviations from the dyadic rules in chapter 4 and by the behavior of bachelors, which resembled the behavior of inhibited rivals in the experiments of Kummer et al. (1974). The inhibitive effect of the father-daughter bond seems to be one of the factors which cause the delayed formation of initial units. Other delaying factors appear to be the effect of the harem bond, which forces the follower to stay at the one-male unit's periphery, and the immature females' tendency to withdraw from play interactions and to groom with their mothers.

This set of triadic rules does not explain all phenomena of family forming. Firstly, the fight between harem leader and follower over the adult females occurs in spite of the inhibitive effect of the harem bond. The occurrence of the fight appears understandable, if viewed in terms of dominance status. It seems that the harem leader has become a less efficient fighter by the time he is defeated by his follower. After all, he loses the fight and, therewith, also his females. If, according to Kummer et al. (1974), there is an inhibition of the follower which reduces his tendency to fight over the harem females, this inhibition is, in the long run, overridden by a marked superiority of the follower. This was already suggested by the experiments of Kummer et al. (1974). The effect of inhibition appears ideal to reduce constant competition over females at short sight, whereas the eventual breakthrough of the dominance principle ensures the fighting capability of harem leaders in the long run.

The second unexplained aspect concerns, I think, an essential problem of hamadryas society. On the one hand, all male-female bonds have a suppressing effect on male-male interactions, which already have the lowest a priori compatibility. On the other hand, the functioning of the hamadryas society depends largely on the relationships among males (Kummer 1968a, pp. 152–3). This discrepancy also appears in family forming: On the one hand, the relationship between harem leader and follower is negatively affected by the harem bond, the father-daughter bond, and later by the initial bond. On the other hand, females who are transferred during the forming of initial units during the breakup of one-male units are preferentially taken over by followers of their former leaders (tab. 25); family forming, therefore, appears to be bound to the leader-follower association. So far we have found that the only factor which appears to promote the development of initial units is the suppression of father-daughter interactions by the harem bond; this may weaken the inhibiting effect of the father-daughter pair gestalt and thus permit at least some interactions between followers and immature females. This effect, however, works in the same way for all bachelors and does not explain why it is the follower who finally succeeds in acquiring females from a harem leader. The followers of band III and of band I who remain attached to their defeated leaders suggest that the bond which attaches a follower to a one-male unit is not only or necessarily the follower's interest for females of that unit. We hypothesize, therefore, that there is a particular relationship between harem leader and follower which might counteract to some degree the suppressive effects of male-female bonds. This could explain why followers acquire their females mainly from their own harem leaders and not from any strangers, to whom they are less closely attached. In order to further our understanding of the leader-follower relationship, we have to turn to the interactions among males from an ontogenetical viewpoint.

7   The Ontogeny of Male-Male Interactions

This chapter deals with the ontogeny of male-male interactions and particularly attempts to reconstruct the ontogenetical development of the leader-follower relationship. We shall at first focus on a characteristic behavior among males, notifying. Until now this behavior has been insufficiently investigated. Our analysis will demonstrate the rivalry for females which exists among the males of the band and which underlies the processes of family forming. We shall then present some results on the leader-follower relationship on the foraging march. Finally, we shall attempt to find the ontogenetical origin of the leader-follower relationship. This will lead us back to family forming, now enriched with the ontogenetical background of male-male interactions.

A. Notifying at the Sleeping Cliff

In the characteristic form of notifying, the actor slowly approaches a sitting partner, looks at him, suddenly turns, shortly presents his anal field and retreats immediately and hastily (Kummer 1968a, p. 128). Initially, the occurrence of notifying seemed to be correlated with changes in relative location among males (Kummer 1968a). A more precise aspect emerged in the inhibition experiments of Kummer et al. (1974): in a triad of a possessor, his female and a rival, it is the possessor who notifies his rival rather than vice versa. In their work, Kummer et al. included this behavior under "presenting," but actually most occurrences correspond to the special form of notifying (Kummer, personal communication). Our observations on family forming suggested that there is a rivalry for females among mature males. We assumed that the inhibiting effect of male-female bonds temporarily prevents the rivalry from taking the form of overt fighting over females. Therefore, we approach our data on notifying with the hypothesis that notifying at the sleeping cliff occurs in the same context as in the inhibition experiments. In particular, we should expect that harem leaders notify mainly their own followers, of whom they have mostly to fear an attack. We also should expect that notifying occurs more often among initial leaders than among harem leaders, because harem leaders were found to be less competitive over females than initial leaders (section 6D).

144

Our analysis is based on the event unit sample which was taken on band I (section E). The males of band I were found to notify 403 times within the band; 36 times they notified males of other bands, and 29 times they were notified by males of other bands. Thus more than 90% of the notifications occurred within the band although males of 2 other bands were available at the sleeping cliff. We shall limit our analysis to the notifications within the band. The events of family forming influenced the frequency of notifying: During the first 3 months of the sampling period, from September until November 1971, 61 notifications were recorded; during the 4 months preceding the breakup of Padorn's one-male unit 125 notifications were observed, and during the remaining 2.5 months after the breakup 217 notifications were registered. The breakup of Padorn's one-male unit will permit us to analyze in more detail some changes of notifying which seem related to rivalry for females; the same analysis could not be made for the breakup of Admiral's and Rosso's one-male units because it took place after the main study period.

The mature males of band I were classified as harem leaders (Padorn, Admiral, Rosso, Bruno, and Bishop), initial leaders (Spot, Cadet, Rossini, and Freund) and subadult bachelors (Stupsie, Coci, Pepsi, and Minimus). Table 27 gives the frequencies of notifying for these 3 classes of males during the whole sampling period. The table shows that most notifications were directed by harem and initial leaders to initial leaders: They represented 72% of all notifications but were concentrated in 21% of all possible actor-recipient combinations ($X^2 = 39.79$, $p < 0.001$). We shall focus on this majority of the notifications.

1. Notifying among Initial Leaders

Notifying among initial leaders for 39% of all notifications and was concentrated in 8% of all possible actor-recipient combinations ($X^2 = 31.34$, $p < 0.001$). Two attempts of Rossini to take over an initial female from Cadet and Spot, respectively, indicated that notifying among initial leaders might be correlated with rivalry for females. The first attempt of Rossini occurred at the beginning of May 1972:

Cadet sits on the rock and is intensively groomed by one of his initial females. He looks continuously at Rossini who sits a few meters away from him. Cadet repeatedly approaches Rossini and notifies him. The grooming initial female seems anxious to remain near Cadet. This is repeated for more than 45 minutes. Then, suddenly, Cadet lunges at Rossini, chases him away, returns to his place, and starts pumping cheeks and yawning at Rossini. Rossini threatens him in turn. Then Cadet again notifies Rossini.

A similar sequence occurred after Spot's first attempt to defeat Padorn

Table 27. Notifying within Band I at the Sleeping Cliff

		Recipients			
		Harem leaders	Initial leaders	Bachelors	Total
Actors	Harem leaders	13 (51.7)	136 (51.7)	6 (51.7)	155
	Initial leaders	26 (51.7)	156 (31.0)	6 (41.3)	188
	Bachelors	10 (51.7)	27 (41.3)	23 (31.0)	60
	Total	49	319	35	403

Figures represent the number of notifications which were recorded during the whole sampling period from September 1971 until July 1972. Figures in parentheses indicate the expectations from the null hypothesis which is made under the assumption that each male notifies any other male with equal probability. The null hypothesis is rejected with X^2 = 794.45, p < 0.001.

(section 6A). The common pattern consists of an excited leader with a female who is anxious to stay near him and the leader's repeated notifications, and finally threats, toward another male. In both cases it was a challenged possessor who notified a rival. This agrees with our hypothesis.

At the end of May 1972 Rossini succeeded in taking over Spot's initial female, Elizabeth (section 6A). After the takeover the new possessor attempted to notify the former possessor, but Spot was not willing to accept the notification:

28.5.1972: Rossini has taken over Elizabeth from Spot.

29.5.1972: In the morning Rossini attempts several times to approach Spot. From previous experiences with other males we feel that Rossini tries to notify. But Spot withdraws from Rossini, hides behind a bush or stays out of Rossini's sight. When the band starts to leave Cone Rock, Rossini waits for Spot who remains in his place. After waiting one minute, Rossini gives up and follows the band. Spot, finally, also leaves as the last animal of the band.

In the evening Rossini sits on the sleeping ledge of Spot. This is unusual, because Rossini's ordinary sleeping place is about 12 meters away from Spot's. Spot does not approach until Rossini leaves and moves to his own sleeping ledge. Then Spot passes Rossini in a long detour and occupies a ledge other than his usual one, out of Rossini's sight.

30.5.1972: In the morning Rossini again tries several times to approach Spot, who withdraws each time. When the band leaves, Rossini again waits for Spot. Spot, finally, approaches Rossini in order to follow the departing band. He starts to pump cheeks and to yawn and then passes by in a hasty gallop.

31.5.1972: In the morning Rossini succeeds in approaching and notifying Spot. In the evening he notifies again.

Our hypothesis says that at the sleeping cliff, it is the possessor of females who notifies his rivals. Since every initial leader is the possessor of a juvenile female, the hypothesis has to be modified: In a situation with two possessors, notifying will be performed mainly by the male whose possession is more challenged by the presence of the other male. Our data suggest that the following 3 factors might determine the degree of challenge: (1) The degree to which the possession is firmly established. A possessor is the more easily challenged the less firmly his possession is established. (2) The attractiveness of the possession. A possessor is the more easily challenged the more attractive his possession is. (3) The degree to which the rival's motivation to take possession of females has not yet been satisfied.

The figures of notifying among initial leaders are presented in table 28 for the period before and after the breakup of Padorn's one-male unit. In band I Freund, Rossini, Cadet, and Spot represented a gradation of initial leaders of whom Freund's unit was least and Spot's unit the most autonomous (section 53). Table 28 shows that before the breakup the initial leaders notified other initial leaders the less frequently the more autonomous their unit became. This agrees with the presumed effect of the first of the above-mentioned factors. The attractiveness of possession should play no noticeable role in this case, because all initial leaders possessed juvenile, and therefore similarly attractive, females. The only recognizable effect could be that Cadet notified other initial leaders more frequently than did Spot, although Cadet's initial unit did not appear that much less autonomous than Spot's; the explanation could be that Cadet's possession of 2 initial females was more attractive than Spot's single female. The rivals' satisfaction of the possessive motive might

Table 28. Notifying among the Initial Leaders
of Band I before and after the
Breakup of Padorn's One-Male Unit

| | | Recipients | | | | |
		Freund	Rossini	Cadet	Spot	Total
Before Breakup Actors	Freund	—	14	2	4	20
	Rossini	9	—	3	5	17
	Cadet	2	7	—	1	10
	Spot	0	4	0	—	4
	Total	11	25	5	10	51
After Breakup Actors	Freund	—	16	5	1	22
	Rossini	7	—	0	13	20
	Cadet	16	8	—	5	29
	Spot	14	19	1	—	34
	Total	37	43	6	19	105

have slightly influenced the figures if we assume that initial leaders with autonomous units are more satisfied than initial leaders in the separation phase; with this assumption, however, Freund has the less dangerous rivals than Spot and, therefore, the effect should work against the effect of the possession's establishment. The increasing autonomy of the initial bond appears the only one of the 3 factors which correctly describes the decreasing notifying frequency in Freund, Rossini, Cadet, and Spot.

Table 28 reveals that after the breakup of Padorn's one-male unit, Cadet and Spot showed increased frequencies of notifying other initial leaders: Before the breakup they notified other initial leaders 14 times, afterwards they notified 63 times. This increase is significant compared to the change for Freund and Rossini, who notified 37 times before and 42 times after the breakup ($X^2 = 14.55$, $p < 0.001$). Through the breakup of Padorn's unit Spot and Cadet acquired adult females, whereas Rossini and Freund obtained juvenile females (fig. 21). The degree of the possession's establishment cannot account for the increased notifications of Spot and Cadet, because all four initial leaders acquired their new females at the same time. The attractiveness of the possession, however, explains the change, because Spot and Cadet acquired the more attractive possessions than did Freund and Rossini. At the same time we should expect that Spot and Cadet address their increased notifications preferentially to Freund and Rossini, whose motivation of possessing females presumably was less satisfied through the breakup of Padorn's unit. The expectation is supported by the figures of table 28: before the breakup Spot and Cadet notified Freund and Rossini 13 times and were notified 14 times by the latter; after the breakup they notified Freund and Rossini 57 times and were notified 19 times by them. The increase for Spot and Cadet is significant (X^2 = 6.60, p < 0.01). My interpretation, therefore, is the following: after the breakup of Padorn's unit, Spot and Cadet notified other initial leaders more frequently than before because they had acquired the more attractive females; they addressed their notifications to Freund and Rossini because the latters' motivation of possession were less satisfied.

Table 28 also shows that Freund and Rossini notified one another 23 times before the breakup, whereas Spot and Cadet notified one another only once during the same period. After the breakup Freund and Rossini notified one another 23 times, Spot and Cadet 6 times. Thus Freund and Rossini notified one another far more frequently than did Spot and Cadet (X^2 = 28.50, p < 0.001); the difference was not significantly affected by the breakup of Padorn's unit. These results agree with the effects of the three presumed factors: Freund and Rossini had established equally attractive possessions about equally well before and after the breakup; the same also applies to Spot and Cadet. Freund and Rossini, however, were less satisfied both before and after the breakup of Padorn's unit.

Thus, Freund and Rossini were considered as dangerous rivals not only by Spot and Cadet, but also by one another. Apparently, their persistent efforts to separate their initial females from the parental units indicate a high motivation to take possession of the female; at the same time the establishment of the possession is delayed by the leader's possessive claim on the juvenile female and, therefore, becomes realized only after a long period of persistent activity. This may increase the tendency to replace the gradual acquisition by an abrupt takeover of *any* females and thus make the young initial leader momentarily a more dangerous rival than the older initial leader. This situation did not improve with the breakup of Padorn's unit, since the younger initial leaders Freund and Rossini acquired less attractive females than the older initial leaders Spot and Cadet. During the whole sampling period Freund and Rossini received 116 of the 156 notifications among initial leaders, Spot and Cadet only 40 (X^2 = 37.02, p < 0.001).

The frequencies of notifying among initial leaders thus are consistent with the hypothesis that notifying is mainly performed by challenged possessors and addressed to males who can be considered as generalized rivals.

2. Notifying between Harem and Initial Leaders

When we tried to understand notifying as a possessor's behavior toward rivals, we first assumed that initial leaders would notify harem leaders more frequently than vice versa, because harem leaders have to be considered as important challengers of the developing initial bond. The fact that the opposite was true appeared inexplicable. Then the fight between Padorn and Spot added a new aspect to our understanding. Spot's first attempt to defeat Padorn pointed out the follower's rivalry for the adult females of his leader, an aspect which we had overlooked. Now it became possible to hypothesize that a double rivalry exists between harem leader and initial leader: one rivalry for the possession of the initial female, which accounts for the initial leader's notifications, and a more prevailing rivalry for the possession of the more desirable adult females, which causes the harem leader to notify. We shall investigate each side of this double rivalry separately.

The main factor which underlies the presumed rivalry of initial leaders for the adult harem females probably is the high attractiveness of adult females and the low degree to which initial leaders could satisfy their possessive motive; the effect of the degree to which harem bonds were established was probably negligible because during the sampling period all harem leaders had well-established one-male units. That a newly established harem bond increases notifications of the possessor has already been demonstrated for Spot and Cadet. If the harem leader's notifications are correlated with the initial leader's rivalry for the adult females, the following results would be expected:

(a) Harem leaders should notify more frequently initial leaders than other harem leaders, because initial leaders could less satisfy their possessive motivation and therefore are more dangerous rivals than other harem leaders. Actually, initial leaders proved to be the main recipients of females in the breakup of one-male units (tab. 25). The figures of notifying agree with the expectation: During the whole sampling period, harem leaders notified other harem leaders 13 times and initial leaders 136 times (tab. 27). The difference is significant for each harem leader: Padorn ($X^2 = 16.34$, $p < 0.001$); Admiral ($X^2 = 15.70$, $p < 0.001$); Rosso ($X^2 = 46.08$, $p < 0.001$); Bruno ($X^2 = 5.33$, $p < 0.05$) and Bishop ($X^2 = 19.70$, $p < 0.001$). The difference remains significant for Padorn, Admiral, and Rosso if notifications to their own initial leaders are omitted ($X^2 = 16.05$, $p < 0.001$). That means that even an initial leader who is not a follower of a harem leader is to him more dangerous as a generalized rival than another harem leader.

(b) Harem leaders should preferentially notify their own initial leaders, who proved to be the main recipients of adult females in the breakup of one-male units (tab. 25). The expectation is supported by the figures

Table 29. Notifying of Harem Leaders toward Initial Leaders in Band I before and after the Breakup of Padorn's One-Male Unit

		Recipients				
		Freund	Rossini	Cadet	Spot	Total
Before Breakup Actors	Padorn	13	1	0	9	23
	Admiral	0	4	7	2	13
	Rosso	4	20	1	1	26
	Bruno	0	1	0	3	4
	Bishop	2	10	3	6	21
	Total	19	36	11	21	87
After Breakup Actors	Padorn	1	0	0	0	0
	Admiral	0	4	3	1	8
	Rosso	3	20	0	0	23
	Bruno	0	3	1	2	6
	Bishop	4	7	0	0	11
	Total	8	34	4	3	49

Underlined figures indicate notifications which harem leaders addressed to their followers.

of table 29. The preference for the own initial leaders is significant for all harem leaders with initial leaders as followers; Padorn ($X^2 = 20.17$, $p < 0.001$); Admiral ($X^2 = 4.91$, $p < 0.05$) and Rosso ($X^2 = 83.29$, $p < 0.001$).

While the processes of family forming suggested that the older followers were the most dangerous rivals to the old harem leaders (tab. 25), the figures of table 29 indicate that the younger initial leaders Freund and Rossini received more notifications from their harem leaders than did the older initial leaders Spot and Cadet: Before the breakup of Padorn's unit, Freund and Rossini received a total of 33 notifications from their harem leaders, Spot and Cadet only 16 notifications. Two reasons might explain this difference: Firstly, Freund and Rossini, who attempted to separate their initial females from the parental unit, appeared at short sight disturbing rivals. This also agrees with the observation that Freund was repeatedly chased away by his leader, Padorn, and that Rossini experienced severe interventions of his leader, Rosso (section 5B); Cadet and Spot appeared more peaceful, although they were, in the long run, more dangerous rivals to their leaders. Secondly, the possessions of Freund and Rossini were not yet fully established, so that the increased notifications of Padorn and Rosso possibly expressed their continued claim on the initial females.

(c) After the loss of his females the defeated leader Padorn should notify initial leaders less frequently than before; in the ideal case he should not notify any more. In fact, Padorn, who had been notifying initial leaders 23 times before he was defeated, did so only once afterwards. The other harem leaders notified initial leaders 64 times before, and 48 times after the breakup of Padorn's unit. The decrease of Padorn's frequency was significantly greater than that of the other harem leaders ($X^2 = 12.84$, p < 0.001).

(d) After the breakup of Padorn's one-male unit, harem leaders should less frequently notify Spot and Cadet, who have acquired adult females and whose motivation of possessing females was presumably more satisfied now. The figures of table 29 show that harem leaders, except Padorn, notified Spot and Cadet 23 times before and only 7 times after the breakup of Padorn's unit; in contrast, they notified the younger initial leaders, Freund and Rossini, who did not acquire adult females in the breakup, 41 times before and 41 times after the breakup. The decrease of notifying the older initial leaders is significant compared to their notifying the younger initial leaders ($X^2 = 6.38$, p < 0.05). Thus, Spot and Cadet were less frequently notified by harem leaders, as soon as they had acquired adult females; they, in turn, began more frequently to notify the younger initial leaders, who did not acquire adult females.

(e) In the relationship of double rivalry between harem leader and initial leader, the harem leader should notify more frequently than the initial leader, because harem leaders possess adult and therefore more attractive females. The observed frequencies (tabs. 29 and 30) support the expectation. The difference is significant for all harem leader–initial leader associations: Padorn-Spot ($X^2 = 9.00$, p < 0.01); Padorn-Freund ($X^2 = 4.76$, p < 0.05); Admiral-Cadet ($X^2 = 10.00$, p < 0.01) and Rosso-Rossini ($X^2 = 27.22$, p < 0.001).

Thus the notifying frequencies of harem leaders toward initial leaders agree with the hypothesis that the notifications are positively correlated with the initial leaders' presumed rivalry for the harem females. Apparently, the breakup of one-male units is the final outbreak of a rivalry whose overt manifestation was so far prevented by the inhibitive effect of the harem bond.

The discussion of the initial leaders' notifications toward harem lead-

Table 30. Notifying of Initial Leaders toward Harem Leaders in Band I before and after the Breakup of Padorn's One-Male Unit

| | | Recipients | | | | | |
		Padorn	Admiral	Rosso	Bruno	Bishop	Total
Before Breakup Actors	Freund	4	0	2	1	0	7
	Rossini	0	0	4	0	1	5
	Cadet	0	0	1	0	2	3
	Spot	0	0	0	0	0	0
	Total	4	0	7	1	3	15
After Breakup Actors	Freund	1	0	0	2	0	3
	Rossini	0	0	1	0	0	1
	Cadet	2	0	1	1	1	5
	Spot	2	0	0	0	0	2
	Total	5	0	2	3	1	11

Underlined figures indicate notifications which initial leaders addressed as followers to their own harem leaders.

ers, which are presented in table 30, will be less detailed because the figures are small. We expect the following results to occur, if initial leaders notify harem leaders because of the latters' rivalry for the initial females:

(f) Initial leaders should notify harem leaders the less frequently, the more their initial units become autonomous. The observed frequencies before the breakup of Padorn's unit agree with the expectation (tab. 30): Freund, who had the least autonomous unit, notified most frequently harem leaders; Spot, who had the most autonomous unit, notified least frequently. This decrease of frequency parallels the decrease of notifying among initial leaders (tab. 28).

(g) After the breakup of Padorn's one-male unit the older initial leaders should increase their notifications toward harem leaders: they have acquired adult females, a possession which is at the same time attractive and weakly established. The older initial leaders Spot and Cadet notified harem leaders 3 times before and 7 times after the breakup of Padorn's unit; in contrast Freund and Rossini, who did not acquire adult females, notified 12 times before and 4 times after the loss of Padorn's females (tab. 30). The difference of change between older and younger initial leaders is significant ($X^2 = 5.11$, $p < 0.05$). 4 of the 7 notifications which Spot and Cadet performed after the acquisition of adult females were addressed to Padorn, the former possessor of the females, and occurred the day after the breakup of Padorn's unit.

(h) Initial leaders should preferentially notify their own harem leaders, who are most likely to interfere with the development of the initial unit. The expectation is supported for the younger initial leaders Freund and Rossini, who notified 10 times their own and 6 times other harem leaders (tab. 30), whereas the random expectation would be 4 times for the own and 12 times for other harem leaders ($X^2 = 11.25$, $p < 0.001$). Spot and Cadet were not seen to notify their harem leaders before the breakup of Padorn's unit. This agrees with the observation that both of them had autonomous initial units and that their harem leaders were never seen to interfere with their initial bonds. In the case of Cadet, however, it remains to explain why he notified other harem leaders, but not his own, in contrast to Spot, who apparently did not feel challenged by any harem leader. The explanation might be related to the observation that Cadet's leader Admiral was never seen to be notified by any male of Cone Rock troop. The only reason I can imagine is that the size of Admiral's one-male unit, which comprised 9 adult females, made it practically impossible to approach him.

3. An Interpretation of Notifying

Our analysis suggests that the majority of the notifications can be interpreted as a behavior of possessors of females toward rivals. This interpretation, which is consistent with the experimental findings of Kummer et al. (1974), throws a new light on the observation of Kummer (1968a) that harem leaders often notify their followers immediately before moving with their units. We can confirm the observation, but I do not think that notifying in this context serves only to attract the follower's attention to the leader's movement. If we argue in terms of rivalry, we should expect that the leader's position as possessor becomes more challengeable when his unit is on the move; then the leader's attention is diverted from his females and his follower. The assumption agrees with the observation that takeovers of females took place on the foraging march and not at the sleeping cliff (section 6A). From this viewpoint it seems plausible that harem leaders notify their followers just before they focus their attention on the unit's or the band's movement.

Why is it exactly notifying which appears in context of rivalry over females? There are 4 aspects of notifying which seem significant to me and which may be related to one another by the framework of the bonding stage model: (1) Notifying occurs only in mature males, i.e., in animals with a low a priori compatibility. (2) Notifying becomes a common behavior as soon as males are involved in family forming. The analysis of family forming has shown that in the mature close-study males grooming with immature females was negatively correlated to grooming with male partners and positively to notifying among males (section 5C). Thus, notifying becomes a frequent behavior when male-male relationships regress. (3) Morphologically, notifying is a ritualized form of presenting. Presenting is the second stage in the sequence of bond forming after the stage of fighting. Kummer (1975) concludes from his experiments with geladas and hamadryas that frequent presenting and notifying reduce the agonistic motivation in the recipient. Since notifying typically occurs in regressing male-male relationships, it probably serves to prevent the regression from reaching the bonding stage of overt fighting. (4) Notifying is not evenly distributed among males, but specifically performed by challenged possessors and addressed to rivals. This observation suggests that not all males might be equally interested in preventing overt fighting and in maintaining friendly relationships with other males. With this suggestion all our factors which tried to define the differentiation between possessor and rival can be summarized in one single rule: In an encounter of two males whose relationship has undergone regression the notifier is the male who would lose more in case of an overt fight.

The assumption that frequent notifying expresses an attempt to avoid overt fighting adds a new aspect to the formation of initial units. So far our analysis has yielded only one factor which facilitates the formation of initial units: the suppressive effect of the harem bond on father-daughter interactions (section 6D). The results on notifying suggest a second factor: Harem leaders may refrain from overt interferences with the initial bond because, by such interference, they would run the risk of losing even their adult females in a fight with their followers.

B. Notifying on the Foraging March

Band I was observed 15 times during the first hour of the foraging march. In this period 124 notifications were registered within the band. During the same period males of band I notified males of other bands only 5 times and were notified by them only 3 times. Thus, more than 95% of the notifications occurred within the band. Again we limit our analysis to the interactions within the band.

Table 31 summarizes the frequencies of notifying for harem leaders, initial leaders, and bachelors. If we compare the figures on the foraging march with the corresponding frequencies at the sleeping cliff (tab. 27), we receive a 2 × 9 contingency table (2 for the conditions "at the sleeping cliff" and "on the foraging march" and 9 for the actor-recipient

Table 31. Notifying within Band I on the
          Foraging March

	Recipients			
	Harem leaders	Initial leaders	Bachelors	Total
Harem Leaders	11 (15.9)	29 (15.9)	0 (15.9)	40
Initial Leaders	25 (15.9)	43 (9.4)	3 (12.7)	71
Bachelors	3 (15.9)	3 (12.7)	7 (9.5)	13
Total	39	75	10	124

(Actors)

Figures represent the number of notifications
which were recorded on 15 foraging marches
during the first hour. Figures in parentheses indicate the expectations from the
null hypothesis that each male notifies any
other male with equal probability. The null
hypothesis is rejected with $X^2 = 177.49$,
$p < 0.001$.

Table 32. Notifying between Harem and Initial Leaders of Band I
           at the Sleeping Cliff and on the Foraging March

Actors:	H-leaders	I-leaders	H-leaders	I-leaders
Recipients:	own I-leaders	own H-leaders	other I-leaders	other H-leaders
At sleeping cliff	73	12	19	14
On foraging march	7	16	9	9

H = harem; I = initial. The table summarizes the notifications
of Padorn, Admiral, and Rosso toward their own and other initial
leaders, and of Spot, Cadet, Rossini, and Freund toward their
own and other harem leaders. It shows that notifying within
leader-follower associations was not homogeneously distributed
at the sleeping cliff and on the foraging march ($X^2 = 28.98$, $p <
0.001$), in contrast to notifying outside leader-follower
associations ($X^2 = 0.27$, $p > 0.05$).

combinations). The hypothesis of a homogeneous distribution is clearly rejected ($X^2 = 34.28$, $p < 0.001$). In order to identify the actor-recipient combinations that contribute most to the lack of homogeneity, the following gross approximation is used: we test the contingency table for homogeneity by the empirical value of each actor-recipient combination with expected value as calculated from the marginal sums (Pfanzagl 1968, pp. 177–85). The observed value is considered different from the expected value in a homogeneous distribution if it lies outside the confidence limits of the expected value, estimated as $\pm 2 \sqrt{\text{expected value}}$. With this criterion we find that the frequencies on the foraging march differ in 2 respects from the frequencies at the sleeping cliff: (1) Harem leaders notified one another more frequently on the foraging march than at the sleeping cliff. (2) Initial leaders notified harem leaders more frequently on the foraging march than at the sleeping cliff. A closer inspection reveals that harem leaders did not make any difference between their own and other initial leaders on the foraging march ($X^2 = 0.75$, $p < 0.05$); in contrast, initial leaders still notified preferentially their own harem leaders ($X^2 = 29.04$, $p < 0.001$, tab. 32). The main difference between notifying on the foraging march and at the sleeping cliff concerns the leader-follower association: At the sleeping cliff it was the harem leader who notified his initial leader; on the foraging march it was the initial leader who notified his harem leader. Notifying between harem leaders and initial leaders who formed no leader-follower association remained proportionately the same at the sleeping cliff and on the foraging march (tab. 32).

Why does the notifying relationship between the harem leader and his follower change on the foraging march? If we maintain the idea of rivalry for females, our interpretation must be the following: At the sleeping cliff it is the rivalry for the adult females which prevails, whereas on the foraging march the initial bond is more challenged. This interpretation, however, does not make much sense: We have argued that on the foraging march the harem leader's attention is directed to the movement of the band, a situation which offers an increased chance for the follower to approach the adult females. The following observations suggest an addition to the hypothesis of rivalry over females.

1. Notifying and Coordination of Travel

The following record illustrates the context which we consider to be typical for notifying on the foraging march:

21.4.1972; 0645: Band I leaves Cone Rock. Rossini walks far ahead. The one-male unit of Rosso and Rossini's initial female, Rosa, follow in a distance of 20 meters.

0646: Rossini sits down. Rosso and the females approach him. Rossini looks at Rosso, stands up, notifies Rosso and takes the lead of the one-male unit. Rosso stops, looks away from Rossini and scratches. Rossini stops as well, 10 meters ahead of Rosso, who sits down and starts feeding. Rossini also sits down. Rosso scratches, Rossini looks at Rosso and yawns. Rosso yawns and scratches. Rossini scratches and threatens a juvenile. Rosso looks at Rossini and yawns. Rossini approaches his initial female, Rosa, who sits nearby. He sits down and scratches. Rosso yawns.

0650: The one-male unit of Rosso is moving again. Rossini walks in front. He repeatedly looks back to Rosso. Rosso stops. Rossini advances further, stops, sits down, scratches twice and wipes his muzzle. Rosso approaches him. Rossini looks at Rosso, then scratches and yawns. Rosso sits down 4 meters away. Rossini looks at Rosso, approaches and presents to him. Rosso shows an intention to mount Rossini. Rosso and Rossini separate.

0700-0715: Rossini, without Rosa, has left the one-male unit of Rosso (see below).

0715: Rossini returns to the one-male unit of Rosso. Rosso approaches him with Rosa in front of the unit. Rossini walks toward the one-male unit, passes Rosa without paying attention to her, then notifies Rosso and takes the lead of the unit. Rosso and his unit follow him.

0720: Rossini still walks in front of Rosso's one-male unit. Rosso, who walks at the rear of the column, slightly changes the direction. Immediately his females turn to the new direction. Rossini approaches Rosso, notifies him, and runs to the front of the one-male unit, also moving in the new direction.

The record illustrates that notifying on the foraging march seems not to be correlated with rivalry for females. In the above example Rosa roams about freely and is neglected by her initial leader. The record suggests that the follower notifies his leader when he occupies the front position of the moving one-male unit. On the foraging march we repeatedly registered the order of march in different one-male units, although we did not apply a standardized and balanced procedure. We distinguished 3 positions which a follower may occupy: in front, in the rear, or at the side of the one-male unit. Our preliminary figures are summarized in table 33. The figures suggest that initial leaders were found more frequently in the front position than bachelors, and that the frequencies of notifying in followers was distributed homogeneously with the frequency of occupying the front position.

The occupation of the front and rear positions in one-male units corresponds to Kummer's (1968a, pp. 124–28) description of the formation of two-male teams. Kummer also pointed out different functions which are correlated with the positions in the column of travel. The harem leader

Table 33. Notifying of Followers and Their Positions in the
          Marching Order of the One-Male Unit

Position of follower:	Front	Side	Rear	Follower notifies his leader
Older initial leaders	11	5	7	3
Younger initial leaders	22	7	13	13
Older bachelors	7	23	23	3
Younger bachelors	1	10	12	0

The table gives the number of times followers were recorded to travel in front, side, or rear position of their unit and the number of notifications they addressed to their leaders. The table shows that initial leaders more frequently occupied the front position than bachelors ($X^2 = 27.51$, $p < 0.001$), and that the number of notifications was homogeneously distributed with the occupation of the front position ($X^2 = 1.58$, $p > 0.05$).

who walks in front of his females is in a position to initiate the movement and the direction, and since he is the dominant animal, his initiative usually is decisive. In the terminology of Kummer (1968a) the harem leader walking in front performs the I- and the D-functions. This remains unchanged when a bachelor is associated with the one-male unit. According to our own observations, bachelors simply followed their one-male units and did not appear involved in the coordination of travel. The situation changes when the follower grows older and becomes an initial leader. In the front position of the one-male unit the initial leader may initiate the movement and the direction (I-role), while the final decision (D-role) remains with the harem leader (see Kummer 1968a, fig. 55).

Some observations suggest that notifying on the foraging march is correlated with performing the I-role not only within, but also between one-male units. It sometimes happened that harem leaders with marked D-competence did not agree on a common direction of travel of the band. In these situations initial leaders appeared to contribute to the band's cohesion with their I-behavior. This is illustrated in the following examples from several marches:

In the above example of Rosso and Rossini the band is moving in a northeastern direction. Only Padorn and his one-male unit tended to the east from the start. At 0700 the one-male unit of Padorn actually turns to the east and deviates from the rest of the band. Rossini presents to his leader and leaves the one-male unit of Rosso. He runs toward the one-male unit of Padorn which is some 100 meters away from the rest of the band. Rossini approaches Padorn, notifies him, and returns to Rosso. Padorn turns to the north and joins the other one-male units. As soon as all one-male units are together, Padorn again turns to the east. Admiral follows him. Rossini also turns to the east; Rosso follows and then the whole band travels in eastern direction. Soon Rosso turns toward the northeast, the original direction of travel. Rossini turns with him and notifies him. The whole band comes to a stop and rests for half an hour. During this time Padorn faces eastward, then stands up and slowly moves to this direction. First Freund, then Spot follow him. They walk more than 100 meters, but nobody else follows. Slowly they come back while, for the first time today, Freund takes the lead of Padorn's one-male unit. Padorn sits down. Freund moves a few steps toward the rest of the band and sits down as well. A few minutes later he approaches his leader, Padorn, and sits 2 meters away from him. Padorn scratches, looks at the band, then at Freund and again scratches. Freund scratches, approaches Padorn, notifies him and again sits between Padorn and the rest of the band. After a few minutes Freund starts to move toward the other one-male units. Padorn stands up and again walks to the east. Freund follows him. Slowly the other one-male units get up and finally the whole band continues its travel in an eastern direction.

The band travels southwest. Padorn, Freund, and Spot turn to the west. Rossini leaves Rosso's one-male unit, approaches Freund, notifies him, and returns to the rest of the band. Freund then notifies Padorn. Padorn sits down, then gets up and walks back to the other one-male units of the band.

From the beginning Padorn and Freund have traveled in a northeastern direction. The rest of the band follows in a distance of more than 200 meters with Spot in front of the band. While Padorn still walks to the northeast, Rosso turns to the north. Only Spot follows Padorn's direction, whereas the other one-male units take Rosso's direction. Padorn, who still is far ahead of the rest of the band, now turns to the north, rests for a few minutes, and then continues in the northeastern direction. Spot approaches Padorn. Finally, Padorn turns westward and walks toward the rest of the band. Spot has reached the one-male unit of Padorn, notifies Padorn, and takes the lead of the unit. Padorn changes the direction slightly toward the rest of the band. Spot adapts to the change and keeps the lead. Padorn notifies Freund, who walks at the side of the one-male unit, and again changes the direction toward the rest of the band. Spot again adapts to the change. Under the lead of Spot the three males, finally, join the other one-male units of the band.

All these observations make it improbable that rivalry for females is the typical context for notifying on the foraging march; instead, they suggest the original interpretation of Kummer (1968a) that notifying serves to attract the recipients' attention to changes of movement or direction. The question as to whether the situations at the sleeping cliff and on the foraging march have any features in common that could explain the occurrence of this same behavior will be discussed later.

2. Family Forming and Coordination of Travel

Our limited observations suggest that coordination of travel was related to the stages of family-forming:

1. Two male teams seemed to be typical between harem leaders and young initial leaders who were in the phase of separating their females from the parental unit. Twenty of the 23 notifications between harem leaders and their followers occurred between harem leaders and young initial leaders. The full autonomy of the initial unit and the acquisition of new females weakened the coordination with harem leaders: Spot and Cadet moved more independently of their harem leaders than did Freund and Rossini. Freund and Rossini also became more independent after they had acquired new females. Before Rossini took over Elizabeth from Spot, he was observed 19 times on 7 foraging marches to move in close association with Rosso; after the acquisition of Elizabeth he was recorded only twice on 8 foraging marches to move with Rosso. With the acquisition of Elizabeth notifications between Rosso and Rossini dropped from 10 to 4 interactions. Freund was recorded to travel with Padorn 17 times on 4 foraging marches before, and 16 times on 8 foraging marches after the breakup of Padorn's one-male unit. Notifications between Padorn and Freund decreased from 5 to 1.

2. Initial leaders who became more independent of their harem leaders exchanged more notifications with other harem leaders. Before the breakup of Padorn's unit, harem leaders exchanged 13 notifications with their own and 7 notifications with other initial leaders; after the breakup, harem leaders exchanged 10 notifications with their own and 25 notifications with other initial leaders This change of frequency is significant ($X^2 = 6.94$, $p < 0.01$). The hypothesis of rivalry for females does not predict such a change of frequency at the sleeping cliff; and actually the analogous analysis at the sleeping cliff does not produce a significant change ($X^2 = 1.29$, $p > 0.05$).

We may hypothesize that initial leaders contribute first to the coordination of travel within the one-male unit and, later, participate in the coordination on the level of the band.

3. After the breakup of Padorn's one-male unit, Freund and Spot formed a two-male team which differed from the teams described so far. The typical two-male team consists of an old harem leader and a young initial leader, whereas Spot and Freund were both young adult males. When traveling together, Spot and Freund did not occupy differential spatial positions: According to our preliminary counts, Spot and Freund were found 11 times each in front of their marching column and 6 times they traveled side by side. During my short field visits in 1973 and 1974 Spot and Freund still formed a two-male team on the foraging march.

4. Defeated leaders did not lose their D-competence. Kummer (1968a) has already stated that the access to females and the influence on the troop movement are not necessarily correlated in hamadryas males. After the loss of his females Padorn did not form a stable two-male team with either Freund or Spot; therefore, he did not participate any longer in the role distribution of I- and D-behavior within two-male teams. As a defeated leader, Padorn combined both I- and D-competences: on all foraging marches which we followed after the loss of his females, Padorn left the sleeping cliff ahead of the other males; and since the other animals usually followed him, it seemed that he still had a marked D-competence. For about half an hour Padorn would take the lead of his band and then fall back to the rear. During my field visit in 1973 I observed a similar behavior in the defeated leader Admiral.

The following record illustrates a typical D-behavior of Zebra, an old male without females in band III (see also Kummer 1968a, p. 140):

Bands I and II have left Cone Rock in a northeastern direction. Band III is still on the rock and faces northwards. Two pseudopods have formed, pointing to the northeast and the northwest, respectively. During the following minutes the band appears undecided as to the direction of departure. Finally, Zebra, who sits in the center of the band, gets up and leaves toward the northwest. The animals of the northeastern pseudopod come back and the whole band follows Zebra.

On two foraging marches we observed that band III passed the resting band I. Both times Zebra walked in front of his band. The only time we saw a band cross a river, it was band III with Zebra leading the way across the river.

The following observation suggests that old males may also serve as "frontier guards" on the foraging march:

Soon after the departure the animals of band I have come to a stop and feed. In a distance of 20 meters band III passes by. Zebra stops, sits between bands I and III, and faces band I while the animals of his band pass behind him. The defeated leader Padorn of band I moves to the periphery of his band, faces band III, and feeds. Zebra gets up and follows his band. Padorn moves a few steps in the direction of band III, faces it and continues feeding.

The next record shows that even an old male who has joined a band only recently can exert his influence on the band movement:

Padorn has left the sleeping cliff in southwestern direction and sits on a tree. Only Rosso, Rossini, and Bishop follow him, while the other one-male units walk southward. Soon the band comes to a halt. Padorn climbs down the tree and slowly approaches the southern units. After a quarter of an hour the band slowly moves to the south, but then stops again. In the rear of the band is Quadro, a defeated leader from band IV who has joined band I. As the band is still hesitating after 10 more minutes, Quadro moves to the front of the band and leads to the south while repeatedly looking back to band I. After walking 50 meters he is followed by the band, and finally the whole band travels southward.

3. Speculations on Notifying

Our limited observations suggest that notifying on the foraging march occurs most frequently when males perform I-behaviour, i.e., when they take initiatives on the movement, the direction, and possibly on the speed of a traveling band. A similar context is also suggested by the observations of Kummer (1968a, pp. 129 and 134). If this is correct, notifying occurs in two different situations: at the sleeping cliff it occurs in the context of rivalry for females, and on the foraging march it occurs in the context of coordination of travel. Are there any common features in these two situations, which might explain the occurrence of the same behavior? At present an unequivocal answer is not possible. On one hand, one may maintain that both contexts have little in common and that they do not necessarily have to share specific features in order to explain the occurrence of the same behavior. Presenting belongs to the behaviors which originally have evolved in the sexual context and subsequently came to be used in other situations (Wickler 1967). The components of appeasement, which apparently prevails in the notifications at the sleeping cliff, and of arousing attention, which seems characteristic for the notifications on the foraging march, both are already part of the original situation: the presenting of the sexually receptive female. These two components may have been differentiated to be used in two different situations: for bond maintenance at the sleeping cliff and for the cooperative acting out of male-male relationships on the foraging march.

On the other hand, it seems intuitively plausible that the situations at the sleeping cliff and on the foraging march share some specific features: (1) The divergence of intentions among males. Possessors and rivals obviously do not have the same goals with regard to females. On the foraging march males also may differ in their intentions of traveling. Here a divergence of intentions seems most likely when a male initiates a change of movement, direction, or speed. According to our qualitative observations no notifications occurred when the band traveled unanimously in the same direction. Divergence of intentions might negatively affect the compatibility among males, and the resulting risk of aggression might be the comprehensive context for notifying among males. (2) The age correlation of the factors which account for the divergent intentions. On a gross scale the possession of females is positively correlated with age: older males are possessors of adult females; younger males possess juvenile females or still are bachelors. On the foraging march the influence on the band movement seems age-correlated as well: D- and I-competences appear to be qualitatively the same, namely, the capability to obtain the agreement of other males to one's own intentions; D- and I-competences possibly differ only quantitatively in their prospects of success, which are positively correlated with age. (3) The notifier is the male who would lose more if it came to an overt conflict over the divergent intentions. Does this interpretation also apply to notifying on the foraging march? An affirmative answer requires the assumption that a notifier on the foraging march is a male whose influence on the band movement is not well established. The first assumption leads to further assumptions, which are partially supported by our figures, and which can serve as hypotheses for further studies: *(a)* Harem leaders seem not to compete for females at the sleeping cliff (section 6D); in contrast, they may disagree among each other in their intentions of travel. Accordingly, they should notify more frequently among each other on the foraging march than at the sleeping cliff. This agrees with our figures (section 7B). *(b)* Harem leaders can lose their females, but they do not lose their influence on the band movement. At the same time, followers have little to lose as possessors of females, but on the foraging march their influence on the movement always risks to be overcome by the D-males' divergent intentions. This might explain why harem leaders frequently notify their followers at the sleeping cliff, whereas followers frequently notify their harem leaders on the foraging march (tab. 32). *(c)* The influence of a D-male on band movement might decrease with an increasing number of D-males. This would explain why harem leaders increased notifications with new harem leaders. *(d)* A dominant male's initiative is less likely to be disregarded by other males than a young male's initiative, and he therefore needs to notify less in order to assert his intentions. This agrees with our qualitative observation that dominant males usually did not notify other males when they initiated the departure and movement of the band.

Thus, the appeasing effect of notifying might also have been at work on the foraging march. It is possible, however, that this effect was particularly marked during our study period, which was characterized by a considerable rivalry for females, leading finally to the turnover of the old harem leaders.

C. The Clan Structure of the Band

The relationship between leader and follower seems important for family forming and coordination of travel. The ontogeny of relationships among males of different ages in general will now be investigated with the aim of tracing the origin of the leader-follower association.

1. Morphological and Spatial Symptoms of Clan Structure

To which one-male unit does the growing male attach himself as a follower? The apparently low tendency of males to migrate (section 2D) suggests that a hamadryas male is most likely to become a follower in the band into which he has been born. Additional indications permit to be more precise. As soon as we had become familiar with the faces of the mature males in band I, we were struck by the morphological similarity between leaders and followers. We found that the mature males of band I could easily be divided into three morphological groupings:

The first grouping comprised the males Rosso, Bishop, Rossini, and Pepsi (see fig. 18). Their characteristic features were the deep red pigmentation of the face and the silver-gray color of the mantle. We used to name this grouping the *Reds*.

The second grouping consisted of Padorn, Bruno, Spot, Freund, and Stupsie. In contrast to the Reds, these males had brown-red faces and brown-gray mantles. We referred to this grouping as the *Browns*.

The third grouping included Admiral, Cadet, Coci, and Minimus. Like the Browns, these males had brown-gray mantles, but their facial pigmentation was a particular violet-brown color. Moreover, the dorsolateral ridges of their muzzles were characteristically protruding. We named these males the *Violets*.

These groupings, which we named "clans," rested and traveled together within the band. Throughout the study period of 1971/72 the Browns occupied the left part of band I's sleeping area, whereas the Violets occupied the central and the Reds the right part. In the week after the breakup of Padorn's one-male unit, the Reds kept spatially apart from the other clans; they moved to the right periphery of the sleeping area and left an empty space of about 5 meters' width to the other clans. In 1973 and 1974 Cone Rock was not regularly occupied by all three bands any more. This led to a new distribution on the sleeping ledges, which appeared less stable than during 1971/72. But there was still a clear tendency for animals of the same clan to occupy neighboring sleeping ledges. One night, in 1974, the Browns slept alone on Cone Rock, whereas the other clans stayed on a small rock about 3 kilometers away from Cone Rock.

The cohesion within each clan also appeared on the foraging march. The members of the same clan tended to stay together within the band,

Table 34. Interactions within and between Clans of Band I

	Within clans		Between clans	
	Grooming	Nongrooming	Grooming	Nongrooming
Interactions among mature males	30	76	3	52
Interactions between mature and immature males	40	16	3	4
Interactions among immature animals	51	194	7	44

The table gives the number of observation sessions containing at least one occurrence of grooming and nongrooming interactions (play, sexual, and maternal behavior). The preference for interacting within the clan is significant for all actor-recipient combinations ($X^2 > 20.00$, $p < 0.001$).

and sometimes they separated from the band and traveled as autonomous sub-units. Several times the Browns arrived alone at Cone Rock, while the other clans arrived later. Subsequent observations of Sigg and Stolba (personal communication) on Cone Rock troop confirm the conclusion that the clan is an autonomous foraging unit in band I.

2. Social Features of Clan Structure

Social interactions within and between clans were estimated as follows: One morning or evening session at the sleeping cliff was used as one observation interval; we counted in all our records the number of observation intervals containing at least one social interaction between or within clans. The resulting figures are summarized in table 34. The procedure is a rough estimate because the observation sample did not include all individuals with equal probability and because for many interactions the clan membership of the animals could not be determined. Table 34 suggests, however, that social interactions occurred more frequently within than between clans.

Table 35 shows the age distribution of the clan males. Of most age classes each clan included one male only. Such a vertical composition should increase the a priori compatibility among males of the same clan if compared to horizontal relationships. Table 34 shows that the preference for interacting with males of one's own clan is stronger for grooming than for nongrooming interactions ($X^2 = 22.49$, $p < 0.001$). Between 1971 and 1973 sixteen of 40 possible male-male dyads within clans were observed at least once to reach the stage of grooming; in contrast, only 8 of 96 possible cross-clan dyads reached that stage ($X^2 = 12.45$, $p < 0.001$). Of the dyads which did not reach the stage of grooming, 15 dyads within clans and 32 across clans presented or notified at least once ($X^2 = 1.97$, $p > 0.05$). Thus, grooming among males is more likely to occur within than between clans, whereas the lower stage of presenting is more evenly distributed.

Table 35. The Male Composition of the Clans of Band I in April 1972.

Male classes	The Browns	The Violets	The Reds
Old harem leaders	(Padorn)	Admiral	Rosso
Young harem leaders	(Bruno)	——	Bishop
Old initial leaders	Spot	Cadet	——
Young initial leaders	Freund	——	Rossini
Old bachelors	Stupsie	Coci	——
Young bachelors	——	Minimus	Pepsi
Males 3	——	——	——
Males 2	Hajö	Hacky, (Mango)	Bub
Males 1	——	Ishi, (1 male)	Zabli, (1 male)
Males N and B	——	2 males	Reso

The names in parentheses indicate males who disappeared until February 1974.

3. On the Ontogenetic Development of Clan Structure

Tables 21 and 23 have shown that subadult males, before establishing their initial unit, frequently groom with male partners, and that these male-male interactions are suppressed by family forming. The figures of the above section indicate that grooming among males most frequently occurred within the clan. I hypothesize that the grooming relationships among young males are the ontogenetical precursor of the leader-follower association, and that they constitute the skeleton of the clan structure. The following observations support this hypothesis: *(a)* The males of the same clan are morphologically similar (see above). *(b)* Nine of the 35 females who belonged to band I in February 1974 (tab. 4) were found to have been transferred to another clan between August 1971 and February 1974 (tab. 25). During the same period only 1 of the 22 males was found to have changed the clan. Thus, males are more likely to remain in their clan than females ($X^2 = 4.18$, $p < 0.05$). The only male who changed the clan during the observation period until February 1974 later returned to the original clan (see below).

The behavior of the transferred females has suggested that mothers and daughters who were separated from each other tended to join again; in one case a presumed daughter even changed clan in order to rejoin her mother (Section 6B). In contrast, two males whose mothers changed clan did not follow their mothers, but remained in their clan.

One of these males was Ishi, a male 1 from the one-male unit of

Admiral. During the breakup of Admiral's unit Ishi's mother was taken over by Bishop, i.e., she changed from the Violets to the Reds. In 1973 Ishi was still regularly found with the Violets, where he frequently groomed with his father, Admiral. Twice he was seen to groom also with his mother in the one-male unit of Bishop. In 1974 Ishi still was in the clan of Violets and still groomed regularly with Admiral, but was not seen to interact with his mother any more.

The second male was Maybe. He was still an infant when he was transferred with his mother from Admiral's to Bishop's unit, i.e., from the Violets to the Reds. During the field visit in 1973 Maybe was regularly found near his mother in the evening. Besides that, he was occasionally observed with his original clan, the Violets. In 1974 Maybe had returned definitively to his native clan of the Violets (Sigg and Stolba, personal communication).

These observations are consistent with the hypothesis that relationships among males who remain within their native clan constitute the clan structure. The hypothesis is further investigated in Section 7D.

4. Stupsie, A Male Who Changed His Clan

The only male of band I who presumably had changed clan for a long duration was Stupsie, a subadult. The assumption was inferred from the observation that Stupsie's morphological features classified him as a member of the Browns, while he was a follower in the clan of the Violets. The following observations also agree with the assumption that Stupsie was not a native member of the Violets:

1. Stupsie appeared to be weakly integrated in the presumable host clan of the Violets. He groomed in only 0.7% of intervals with mature males; among the close-study bachelors only Pirat from band III was never seen to groom with mature males; the other bachelors' mean interval percentage of grooming with mature males was 15.1% (5.2–20.6%). Even initial leaders, whose interactions with male partners were suppressed by the initial bond, still groomed with mature males with a mean frequency of 2.8% of intervals. Within the clan of the Violets Stupsie groomed only once with Cadet, whereas Cadet, Coci, and Minimus, the other followers of the Violets, regularly groomed one another.

Stupsie's isolation among the mature males of his band seemed to make him an easily accessible partner for newcomers. Two defeated leaders from other bands who joined band I in 1972 and 1974, respectively, groomed exclusively with Stupsie.

2. Stupsie's formation of an initial unit appeared to be delayed. In 1974, when Stupsie already was a young adult, he still had no initial female, in contrast to Coci, Pepsi, and Minimus, who were younger than Stupsie and who all established autonomous initial units during 1974

(Sigg and Stolba, personal communication). Although Stupsie did not interact less frequently with immature females than did the other bachelors in the onset phase (tab. 21), these interactions apparently did not lead to a fully established possession. I do not have any causal explanation for this phenomenon.

The frequencies of notifying among bachelors also illustrate the particular position of Stupsie: He notified other bachelors of his band only once, in contrast to the other bachelors: Coci notified other bachelors 10 times, Pepsi 9 times and Minimus 3 times. According to our results on notifying we may interpret that Stupsie, similarly to Minimus, who hardly had entered the onset phase of family forming, was an unchallenged "possessor," because he had not even a prospective possession as had Coci and Pepsi. On the other hand, Stupsie received 18 notifications from the other bachelors, Coci only 2, Pepsi none and Minimus 3. Nine of the 18 notifications which Stupsie received were performed by Coci, who was his age mate and who also was a follower in the clan of the Violets. Stupsie's advanced age, the retarded development of his initial bond, and his attachment to a strange clan seemed to make him a potential rival to other bachelors.

3. Stupsie attempted to join the Browns, his presumed clan of origin, after the breakup of Padorn's one-male unit. He repeatedly approached the defeated Padorn and presented to him; on the foraging march he was sometimes seen to travel with Padorn (tab. 19). At the same time Stupsie attempted to become a follower of Spot. Spot's intolerance seemed to make a definitive attachment impossible. During the field visit of 1973 Stupsie was occasionally found in the vicinity of Spot, but mostly he stayed at the band's periphery. Finally, in 1974, Stupsie seemed to succeed in becoming a follower of Spot. Several times he was observed to mother Spot's first daughter, a female B, and to groom with her. We expected Stupsie to remain in the clan of the Browns; but at the end of 1974 he disappeared and possibly died (Sigg and Stolba, personal communication).

The features of Stupsie's social situation may not be unique. In band II, 5 of 6 followers physically resembled their leaders. The exception was Rugel. He was about the same age as Stupsie and also was associated with a harem leader whom he did not resemble. By 1974 Rugel was the only male of his age class in band II who still was without an initial female. Stupsie and Rugel suggest that a male may succeed in gradually forming an initial unit only within his native clan.

We can only speculate on the reasons which might have caused Stupsie's presumable change of clan. The number of "open positions" for followers is limited by the number of existing one-male units. The intolerance of harem leaders, the number of bachelor candidates, and possibly the size of one-male units may further limit the opportunities of a would-be follower. If it is true that a male normally becomes a follower within his native clan, there might sometimes be more subadult males in a clan than open positions for followers. Stupsie may be an example of the consequences; there were 2 one-male units in the clan of the Browns. But the old leader Padorn already had 2 followers, Spot and Freund, who both were older than Stupsie. The other one-male unit was that of the young leader Bruno, an excessively intolerant leader who vigorously herded even his two females B. Bruno was never seen to have a regular follower (tab. 19). We assume that Stupsie was a native member of the Browns and that he attached himself to the one-male unit of Admiral only because he had no opportunity to become a follower in his native clan.

D. The Origin of the Leader-Follower Association

1. Age Estimates and Age Classes

If hamadryas males remain within their native clan for life, they are bound to each other by kin relations. In order to investigate possible kin relations we require estimates of actual ages in years. Our observations from the middle of 1971 until the beginning of 1974 permitted us to register changes of age class over 2.5 years for several marked individuals. The results of these changes and the inferred estimates of age are summarized in figure 26 for males and in figure 27 for females. Age estimates for adult males are explained below. The durations of the infant, juvenile, and mature phases are compatible with the values for macaques (Jolly 1972, fig. 100). According to our estimates, sexual maturity is reached at 3 to 4 years in females and at 5 years in males. These estimates agree with the figures for macaques (Itani et al. 1963, Simonds 1965, Spiegel 1954), langurs (Jay 1965) and baboons (Altmann and Altmann 1970, Ransom and Rowell 1972).

This age scale permits us to estimate the age differences within some vertical relationships:

Initial bonds: Band I provided us with several examples to estimate the age difference within initial units: Spot-Elizabeth, 6 to 7 years; Cadet-female 2, 6 to 7 years; Rossini-Rosa, 6 to 7 years; Freund-Freundin, 6 to 7 years; Pepsi-Prima about 6 years. The overall tendency indicates an age difference of 6 to 7 years.

Leader-follower associations: The best approximations available are the age differences within younger leader-follower associations, because the age estimates for older leaders are hardly reliable. Spot-Stupsie, 3 to 4 years;

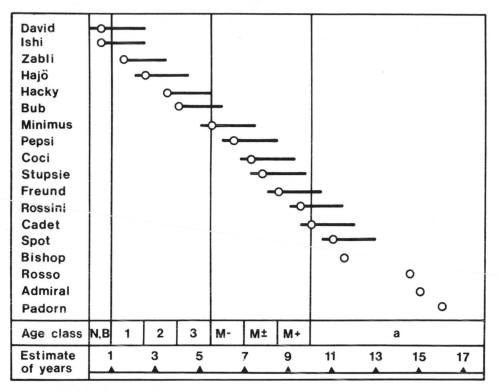

	N,B	1	2	3	M-	M±	M+	a
David								
Ishi								
Zabli								
Hajö								
Hacky								
Bub								
Minimus								
Pepsi								
Coci								
Stupsie								
Freund								
Rossini								
Cadet								
Spot								
Bishop								
Rosso								
Admiral								
Padorn								
Age class	N,B	1	2	3	M-	M±	M+	a
Estimate of years	1	3	5	7	9	11	13	15	17

Figure 26. Age estimates for males. The circles indicate the age class of marked individuals at the beginnig of 1972. Horizontal lines on the right side of the circles represent the change of age class until the beginning of 1974; lines on the left side mark changes of age class since the middle of 1971. The estimates of years are based on these changes of age class. For the estimates of adult males see section 7D.

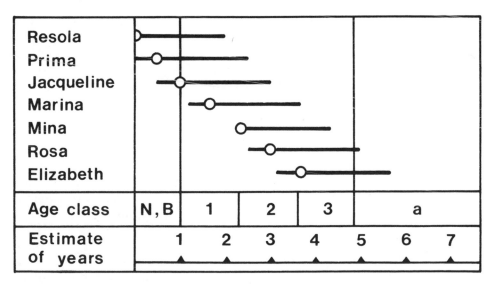

	N,B	1	2	3	a
Resola					
Prima					
Jacqueline					
Marina					
Mina					
Rosa					
Elizabeth					
Age class	N,B	1	2	3	a
Estimate of years	1	2	3	4	5	6	7

Figure 27. Age estimates for females. For explanation see figure 26.

Cadet-Minimus, 4 to 5 years; Bishop-Pepsi about 5 years; Polo-Gox (band II) about 5 years. The central tendency is about 4 to 5 years. If this is correct, we may infer that an old harem leader is about 16 years old when he is defeated by his oldest follower. Since the defeated leader may survive the fight, a hamadryas male can presumably reach an age of more than 20 years in the field.

Father-son relationships: According to these estimates the males Bishop, Spot, Cadet, and Freund were about 10 to 12 years old when they had their first infant. Even if our age estimates are not very precise, this difference of age excludes the possibility that followers are the sons of their leaders.

2. A Hypothesis on the Ontogenetic Origin of Vertical Relationships

The review of some previous results in the light of the age estimates permits us to present a hypothesis on the ontogenetic origin of vertical male-male and male-female relationships. When describing the ontogenetic development of the initial bond, we found that maternal behavior toward female infants was typical for the onset phase (tab. 21). Maternal behavior toward infants usually occurred in interaction bouts which Kummer (1968a) described as "kidnapping" (fig. 28). The hypothesis which is presented here says that kidnapping infants not only is the characteristic onset behavior of family forming, but also that it is the ontogenetic origin of the leader-follower association. Since we could not identify enough immature males in band I, we cannot describe the development of single male-male dyads as we could do it for family forming. Instead, indirect evidence is provided to make the hypothesis at least possible:

If the age difference in initial units is estimated to be 6 to 7 years, we should expect that kidnapping female infants is most frequent in 6- to 7-year-old males, i.e., mainly in males M-. If the leader-follower association has its origin in kidnapping male infants, we should further expect that the kidnapping rate is highest in 4- to 5-year-old males, i.e., in males 3. To test this hypothesis we collected all incidents of kidnapping black infants included in our behavioral records. The resulting sample consists of 139 incidents of kidnapping male infants and of 129 incidents of kidnapping female infants. Kidnapping rates are defined as the number of kidnappings per number of available kidnappers in a given age class (see tab. 3). The kidnapping rates are presented in figure 29. The graph shows that kidnapping female infants occurred most frequently in males M− and M±, who are estimated to be from 5 to 9 years old, whereas kidnapping male infants was most frequent in males 3. Thus male infants tend to be kidnapped by younger males than do female infants.

Figure 29 also shows the rate of kidnapping infants of either sex which were accompanied by yawning of the kidnapper. The frequency parallels the distribution of kidnapping females. We may remember that immature females tend to be herded by their harem leaders; a kidnapper who approaches a female infant enters a situation of rivalry with the infant's leader. The gradual development of initial units takes place within the clan, and the leader-follower association is part of the age hierarchy of males, which constitutes the clan structure. Therefore, kidnapping infants should take place preferentially within the clan. One hundred nine of 128 observed kidnappings in band I did in fact occur among members of the same clan. Taking possible errors of the age estimates into consideration, the findings on kidnapping infants appear to be compatible with the hypothesis mentioned above.

Kidnapping interactions between infants and juvenile or subadult males are supposed to be transformed into regular grooming relationships. In 13 close-study animals whose records contained regular grooming between immature females and mature males the mean age difference between the grooming partners was 6 ± 1 years. This is compatible with the estimated age difference for initial bonds. That these grooming interactions appear to be a continuation of the kidnapping interactions and that they tended to occur within clans was described in chapter 5. In the close-study males we found a mean age difference of 4.5 ± 1.5 years (n = 23) for grooming between immature and mature males. This agrees with the age difference which we estimated for leader-follower associations. That these grooming interactions preferentially occurred within clans was shown in table 34.

These observations agree with but do not prove the hypothesis that kidnapping female infants is the ontogenetic onset of family forming, whereas kidnapping male infants is the onset of the leader-follower association. Boese (1975) has independently suggested the same ontogenetic origin for the leader-follower association in Guinea baboons. The similarity in the developments of the leader-follower bond and the initial bond suggests to use a common term: I propose to refer to the precursor of the leader-follower association as the *male-male (MM-) initial bond*, in analogy to the *male-female (MF-) inital bond*.

The estimates of age add another aspect to the hypothesis on the ontogenetic origin of MM-initial bonds: The mean age difference between males of adjacent age classes within clans was 2.5 ± 1 years (n = 17). This is half the estimated age difference in leader-follower associations. It suggests that males immediately adjacent to one another in the quasilinear age hierarchy of the clan are unlikely to form a leader-follower association. In chapter 8 I shall propose that maternal behavior is characteristic for highly compatible, and play behavior for less compatible dyads. Thus, males of adjacent age classes, who possibly become

Figures 28a and b. **Kidnapping infants by juvenile and subadult males. Kidnappings are interaction bouts, that last for a few minutes, and that are characterized by maternal behavior toward infants: invitation to carry, carrying** *(a)*, **embracing** *(b)*, **and grabbing.**

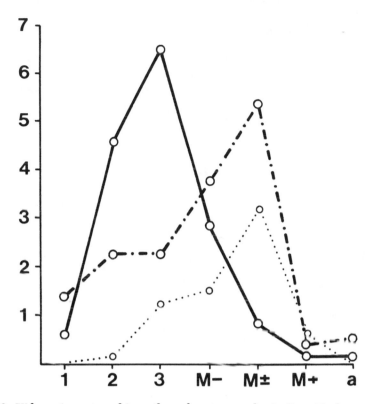

Figure 29. Kidnapping rates of juvenile and mature males in Cone Rock troop. For calculation of rates see text. The graph shows the rate of kidnapping male (———) and female (—·—·) infants, and the rate of kidnappings that were accompanied by yawning of the kidnapper (·········). The hypothesis that kidnapping male and female infants are homogeneously distributed across age classes of kidnappers is rejected with $X^2 = 37.70$, $p < 0.001$.

familiar with one another in play groups, seem not to be compatible enough for a leader-follower association, so that leader-follower associations occur preferentially between males who are separated by a male of intermediate age. This preference was actually observed after the turn-over among the harem leaders in band I (section 6B): Cadet's newly established one-male unit was not followed by Coci, who succeeds Cadet in the age hierarchy of the Violets (tab. 35), but by Minimus. Rossini's enlarged unit was followed by Pepsi; Pepsi was the next male after Rossini in the hierarchy of the Reds and his association with Rossini proved temporary. According to the above-mentioned rule, we should expect that Bub becomes a follower of Rossini; in fact, Bub attached himself to Rossini's one-male unit in summer 1974 and acquired Rossini's initial female, Rosa, in spring 1975 (Sigg and Stolba, personal communication). The new one-male unit of Spot, finally, was followed by Stupsie. In the age hierarchy of the Browns the males who succeeded Spot were Freund and then Stupsie. That Stupsie's attachment to Spot's unit took place slowly, may have been caused by Stupsie's previous stay in the clan of the Violets.

Since observations on Cone Rock troop are being continued, we may hope that these hypotheses will eventually come to the test.

3. Chance Observations on Vertical Male-Male Relationships

The following chance observations illustrate some aspects which may be typical for vertical male-male relationships:

1. *Intervention of older followers in kidnapping interactions:* Harem leaders seemed more likely to intervene in interactions between subadult males and immature females than in interactions between subadult and immature males (section 5B). As a consequence, kidnapped male infants sometimes could not return to their mother because the mother did not dare to approach the kidnapper while the harem leader showed no intention to intervene. In one case a kidnapper prevented a male infant for 30 minutes from returning to his mother. The following observations illustrate that in such situations older followers may intervene and induce the release of the male infant:

> A male infant attempts to return to his mother, who sits with her unit 3 meters above him. He is constantly blocked by a male 3, who interposes between infant and mother. The mother screams, but her leader gets up only once, looks down at the male 3 and pays no further attention to the situation, which remains unchanged for several minutes. Suddenly the one-male unit's subadult follower, who had been seated above the unit, appears, stops 2.5 meters above the kidnapper, and looks at him. The latter looks away as if he were not involved. In this moment the mother runs at her infant and grabs it quickly.

> Minimus, a male M- and the youngest follower of Admiral, has kidnapped a male infant of Admiral's one-male unit and carries it away on his belly. The mother of the infant follows Minimus, looking back to her leader. Minimus sits down and embraces the infant. The mother approaches Minimus and invites her infant to be carried (a behavior which at the same time looks like presenting to Minimus). As the infant cannot get away from Minimus, the mother tries to grab it, but Minimus bites her on the neck. While the mother screams and again tries to grab her infant, Coci, the older follower of Admiral, appears and chases Minimus away. Only now the leader Admiral responds to the situation: he approaches the female and bites her on the neck!

2. *Similarities between MF- and MM-initial bonds:* Subadult males sometimes tended to keep immature males in their vicinity in a way that resembled the mild herding in MF-initial bonds:

A subadult male approached a male 1, who presents. The subadult touches the juvenile's genitals and is then groomed by the juvenile. As the juvenile stops grooming and walks away, the subadult follows, grabs and embraces him, while yawning at the same time. The male 1 grooms again. Then the subadult walks away, sits in 2 meters and looks back at the juvenile, who approaches and grooms the subadult. Again the subadult walks away, looking back at the male 1, who runs away. The subadult chases after him, grabs and embraces him, while pumping cheeks and yawning. The juvenile grooms. As he tries to leave, the subadult again grabs and embraces him while yawning. The juvenile grooms. Finally, the male 1 leaves without interference of the subadult. The whole sequence lasts about 10 minutes.

Stupsie sits together with a male 2. Spot and his female approach the 2 males. Stupsie leaves, repeatedly looking back at the male 2, as a leader does with his female. The juvenile follows Stupsie.

3. *Triadic differentiation in male-male interactions:* The bachelors of band III frequently joined one another in grooming parties. These males, who were not involved in establishing initial units (tab 21), provided a good opportunity to observe the ontogenetic continuation of vertical male-male relationships under conditions that were not affected by family forming. Unfortunately we did not take full advantage of the opportunity, because we did not recognize its importance during the field period. One theoretically important characteristic of these groupings was the type of triadic differentiation found by Kummer et al. (1974) among two males and one female. The following scene involves the young adult subadult male Garu, who appears in a similar position as the female in the above-mentioned experiments:

Steward approaches Garu, followed by Shadow. Steward embraces Garu. Shadow sits behind Steward and starts grooming him. Suddenly he hits Garu, who runs away and screams, while Shadow chases after him. Steward lunges at Shadow and chases him away. All 3 males disappear from sight. As I discover them again, Steward and Shadow are jointly threatening some far-sitting animals of their band (redirected aggression). Then Steward approaches Garu, mounts and grooms him. Shadow has followed and sits 2 meters away from the grooming pair. For more than 15 minutes Garu and Steward groom one another alternately. Garu grooms as intensely as a female which has been neck-bitten by her leader. During this period Shadow performs all the behavior of an inhibited rival: he scratches, grooms himself, wipes his muzzle, sits with lowered head, or lies on his belly. Finally, Steward approaches him, shows an intention to notify him, and sits 1 meter away from him. For a few minutes Steward remains seated between Shadow and Garu and shows frequent conflict behavior, while Garu grooms himself. The situation changes as Garu walks away and starts grooming with another subadult male. In this moment Steward approaches Shadow, embraces and grooms him.

Steward and Shadow normally were closely associated with each other. They not only groomed frequently, but embraced each other when meeting after a separation. Yet, the triadic situation induced a polarization of the relationships, with Garu as "female," Steward as "possessor," and Shadow as "rival"; the relationship of Steward and Shadow underwent a regression from which it recovered only after the removal of Garu. When Steward finally acquired a juvenile female, he was not seen any more to groom with Shadow, but he still groomed with Garu. These observations agree with Kummer's (1975) rule for geladas that the less compatible a dyad is in isolation, the greater is its tendency to regress in the presence of others. Accordingly, they also agree with the viewpoint that only males who differ sufficiently in age can maintain their relationship in the group context, particularly in the context of female possession. This might be an explanation for the quasilinear age hierarchy within clans of band I and for the considerable age difference of 4 to 5 years in leader-follower associations.

E. Discussion

Chapters 5 and 6 left us with the open question why families are formed in spite of the high dominance status of harem leaders and of the inhibiting effect of harem bonds. The results of this chapter offer male-male relationships which are particularly strengthened as a possible solution. It seems that various factors are combined in the hamadryas organization in order to strengthen male-male relationships: (1) The males of a band form quasilinear age hierarchies which constitute the skeleton of a new social unit, the clan. The vertical male composition within the clan appears to be suitable to increase the a priori compatibilities among males, compared to a horizontal composition of males of equal age. Eisenberg et al. (1972) refer to social groups with such an age hierarchy of males as "age-graded male troops." The stabilizing effect of quasilinear age hierarchies is also suggested for langurs: According to Yoshiba (1968), male-male relationships appear to be more relaxed and peaceful in the age-graded male troops at Dharwar than in the multimale troops at Orcha and Kaukori. Furuya (1968) reports that for Japanese monkeys small age differences among leaders and subleaders are correlated with social instability and that thereafter the emigration of males and the fission of troops produce sufficient age differences in the leader class and again lead to social stability. (2) Maternal behavior of bachelors toward male infants seems to be the typical onset behavior of

MM-initial bonds, which are assumed to persist as leader-follower associations. In the next chapter maternal behavior will be interpreted as the
highest bonding stage, which occurs only in highly compatible dyads.
The high compatibility between bachelors and male infants appears to
be an important factor for the strengthening of vertical male-male relationships. (3) The ontogenetic onset of MM-initial bonds seems to pre-
cede family forming. MM-initial bonds may only be possible because the
more compatible MF-initial bonds are themselves delayed. I wish to
discuss this aspect furthermore in the following chapter. (4) Presenting
appears to reduce the antagonistic motivation in the recipient (Kummer
1975). At the sleeping cliff it occurred preferentially when male-male
relationships underwent regression and thereby risked reaching the stage
of overt fighting. The important function of presenting among males is
emphasized by the optical accentuation which this behavior has received
in hamadryas compared to other baboons: enlargement and red pigmentation of the anal field, and ritualization of the movement (notifying).

The clan structure seems to be the result of the strengthening of male-
male relationships. The clan can be defined as an association of a few
family units which differ from each other in their stage of development
and which are linked through a quasilinear age hierarchy of males. Since
males appear to remain within their native clan at least until the age of
reproduction, kin relations may be the basic structure of the clan. The
clan appears to be a social unit whose structure is determined patrilineally.

The vertical age composition of males from the same clan appears to
be accompanied by a functional differentiation. Three functional types
of males can be distinguished: (1) dominant males who are leaders of
reproductive units and decision makers on the foraging march; (2) subdominant males who are leaders of nonreproductive units and who take the
initiatives on the foraging march; (3) subordinate males who possess no
females and who probably are the sons of the dominant males and become the followers of the subdominant males. A fourth type is represented by defeated leaders. The normal transition from the subdominant
to the dominant stage seems to take place through fights between subdominant and dominant males. The observations in band I suggest that the
replacement of the dominant males can be more or less synchronized
across clans; from estimates of age we may infer that the turnovers occur
in intervals of about 5 years. If this is correct, we should expect that the
presently dominant males, Spot, Cadet, and Bishop, would be defeated
in 1977 or 1978. A similar interval in the turnover of dominant leaders
was found by Sugiyama (1965b) in hanuman langurs, whereas Rudran
(1973) estimated an interval of about 3 years for purple-faced langurs.

The replacement of dominant males leads to a shift in the functional
differentiation among the males: subdominant males become dominant

leaders, subordinate males become followers of the new one-male units and start to establish their own family unit. To become a follower appears to be difficult, possibly because of the intolerance of the new harem leaders. This agrees also with Kummer's (1968a, p. 54) observation that initial units usually have no followers. Defeated males, who have reached the postreproductive stage, may help to maintain the cohesion of the clan in that they offer an opportunity for social interactions to males who cannot easily associate with a new one-male unit : In the clan of the Browns Stupsie and Hajö associated with the defeated Padorn until they became followers of Spot. In the clan of the Violets Coci, Hacky, and Ishi stayed regularly near the defeated Admiral until Coci established his own initial unit, and until Hacky became a follower of Cadet and Ishi a follower of Coci (Sigg and Stolba, personal communication). In the clan of the Reds, finally, Bub remained near his father, Rosso; then he became a follower first of Bishop and later of Rossini.

8   A Framework for Socialization

"You'll reply that reality hasn't the least
obligation to be interesting. And I'll answer you that reality may avoid that obligation but that hypotheses may not."
(Borges 1964)

A. The General Outlook

This final chapter attempts to interrelate our results and to develop a
consistent conceptual framework for socialization. Since I wish to emphasize a synthetic viewpoint, I shall not refrain from simplifications and
generalizations. The following theoretical statements will serve as a
guide: (1) Socialization is a network of bond forming which attaches
individuals in varying degrees to one another. The whole pattern of
varying degrees of attachment constitutes the social organization. This
broad definition intentionally goes beyond the mere development of
individual life cycles, for which the term "individuation" would seem to
be more appropriate. (2) Bond forming is a process leading from unfamiliarity to familiarity. Familiarity is irreversible. (3) From a long-term
viewpoint two interrelated functions of social organization should be
taken into consideration: the long-term identity of social units and the
genetic exchange between units.

The bonding-stage model appears to be a suitable starting point to
develop a conceptual framework, because, as Kummer (1975) has already pointed out, its possible merit is not that it describes components
of social structure, but that it may provide a nexus among them. The
concept suggests to distinguish three elements: (1) The stage sequence of
bond forming which permits establishment and maintenance of social
relationships on several bonding stages. This sequence allows varying
degrees of attachment along a single dimension. (2) The sex- and status-
dependent propensities in bond forming which provide a primary pattern of dyadic or a priori compatibilities. (3) The triadic differentiations
which modify the primary pattern of a priori compatibilities.

For the subsequent sections I propose the following procedure: the
description of a generalized stem structure of primate society will relate

the bonding-stage model to social organization. The resulting framework will permit presentation of the particularities of hamadryas organization and, therewith, of hamadryas socialization. The discussion of some particular problems and the presentation of some final speculations will conclude the chapter.

B. The Stem Structure of Subhuman Primate Society

The term has been introduced by Chance and Jolly (1970) in order to describe some general features of primate, particularly cercopithecid, social organization. The authors distinguish three elements within the stem structure: the assembly of females, the cohort of males, and the cluster of juveniles. The terms are used here to achieve a preliminary synthesis of social organization and the bonding-stage model.

(a) The assembly of females: In the stem structure of primate, if not mammalian, society the long-term identity of social units seems to be warranted by the assembly of females. From the viewpoint of the bonding-stage model the assembly of females appears most qualified for this function: (1) Females have a high a priori compatibility for horizontal relationships; this makes them suitable members of a stable and permanent stock within social units. (2) The mother-offspring bond is the paradigm of a vertical, highly compatible relationship and provides the starting point for integrating the next generation into the assembly of females. Thus, the typical long-term structures of the assembly of females are the female lineages as they have been found in rhesus and Japanese monkeys (Sade 1965, Kawamura 1965).

(b) The cohort of males: While females tend to remain within their native groups, males tend to leave them and thus to contribute to the genetical exchange between groups. In Japanese monkeys Kawanaka (1973) observed that within 5 years 104 of 202 identified males left their troop; Koyama (1970) reported that mature males tend to follow their mother when troops are divided but then change to the other part of the divided troop. Drickamer and Vessey (1973) found that in a rhesus colony of Puerto Rico at the age of 7 years all males had left their native troop. Rhesus as well as Japanese males rarely return to their original troop (Drickamer and Vessey 1973, Kawanaka 1973).

The following factors may promote the males' leaving: (1) The low a priori compatibility among males may drive some males to the troop's periphery and thus induce their leaving (Nishida 1966), Simonds 1973). (2) The inhibiting effect of male-female associations on bachelors. In rhesus and Japanese monkeys, males leave the troop most frequently during the mating season (Boelkins and Wilson 1972, Kawanaka 1973); in hanuman langurs, males who are intolerant toward other males in the heterosexual group may live in all-male groups when defeated as leaders of the heterosexual group (Sugiyama 1965b). (3) The influence of the assembly of females. In langurs (Jay 1965) and Japanese monkeys (Yamada 1971) adult females may chase younger males away from the troop's core. Vessey (1971) reported for a rhesus group that the second-ranking male was prevented by the adult females from entering the core of the group after the alpha male had been removed. In the extreme case this may lead to troops without mature males in the central part, as it was observed in Japanese and rhesus monkeys (Kawamura 1965, Neville 1968).

(c) The male-female relationships: In the stem structure of primate society males and females usually form consort relationships. The temporary bond between sexes seems advantageous to both the assembly of females and the cohort of males: The assembly of females does not disintegrate; females can freely associate and establish their lineage structures. In the cohort of males the low a priori compatibility among males is not further suppressed by permanent male-female bonds. From the bonding-stage model we should expect that male-male relationships regress during the mating season and recover during the nonmating season. The expectation agrees with the observations in Japanese monkeys that males groom and mount one another less frequently during the mating season (Hanby and Brown 1974, Tokuda 1962).

(d) The cluster of juveniles: Sex specific differences of behavior occur already in immatures and show a general tendency in primates: Males tend to become independent of their mothers at an earlier age and to play more frequently and more vigorously than do females. From the viewpoint of the bonding-stage model these behavioral differences can be interpreted as differences in a priori compatibilities. The primary pattern of a priori compatibilities appears to be differentiated at an early stage of ontogeny.

I propose to relate high compatibility in horizontal relationships to the long-term function of group identity, low compatibility to the function of intergroup exchange.

C. The Hamadryas Society

The outlines of a generalized primate society may help us to recognize some particular features of the hamadryas society. The main difference from the stem structure of primate organization concerns the male-female relationships, which became permanent harem bonds in hamadryas. Kummer (1968a) has emphasized the evolution of male herding behavior as one major change toward hamadryas society. Our study suggests that further changes had to take place. From the conceptual framework which describes the stem structure of primate organization we may expect two major problems to arise with the evolution of permanent harem bonds: (1) Permanent harem bonds, particularly if based on herding behavior of males, disturb the network of the assembly of females and, herewith, the long-term identity of social units. (2) Permanent harem bonds permanently suppress male-male relationships. There is no longer a nonbreeding season which would permit the males to interact more or less on dyadic terms with one another.

Our observations suggest that both problems have found one common solution: the strengthening of male-male relationships. Above I have already summarized the factors which appear to contribute to the strengthening of male-male relationships: optical accentuation of presenting; ontogenetic precedence of vertical male-male relationships to male-female relationships in family forming; use of vertical relationships with their high compatibility in clan forming; use of mother-infant behavior as the ontogenetic onset. The strengthening of male-male relationships leads to three features which seem typical for hamadryas organization: (1) It reduces the tendency of males to leave their native group. Male transfers between bands occurred rarely and involved males who were unlikely to mate in the new band (section 21). (2) The ontogenetic development of male-male relationships leads to a quasilinear age hierarchy which seems to be based on kin relations and which forms the skeleton of a new social unit, the clan. The typical feature of hamadryas organization is not the persistence of kin relations among males as such. In rhesus monkeys, as well, brothers may transfer to the same new troop (Kaufmann 1965; Koford 1966), but their kin relation forms only part of the cohort of males in the new troop. In hamadryas, the low tendency of males to emigrate permits kindred males to establish a pervading kin structure that makes up the whole male cohort of their native group. (3) The strengthening of male-male relationships may account for the regular grooming interactions among adult hamadryas males, in contrast to cynocephalus males, who rarely groom one another (Kummer 1968b).

An overall description of hamadryas organization may tentatively distinguish two systems which contribute to the long-term identity of social units: the phylogenetically ancient female lineage system, which has to some degree been fragmented by permanent harem bonds, and the male clan system, which is superimposed on the assembly of females and compensates for its fragmentation. This tentative description assumes that features of the female lineage system still should be recognizable in the hamadryas society; it makes us aware of the fact that until now the long-term function of female-female relationships in hamadryas has largely been neglected. So far the assumption of a persisting female lineage system is supported by two kinds of evidence: Firstly, females attempt to maintain relationships with familiar females and develop regular grooming interactions with unfamiliar females rather slowly (section 6B); secondly, family forming tends to keep females within the clan (tab. 25). The latter observation is a precondition if the alleged female lineage system is to persist at all. Female lineage system and male clan system generally will work to the same effect: a young male who becomes integrated into the clan of his native one-male unit is usually not separated from his mother. Besides familiarity among males of the same clan, the probable persistence of mother-son bonds (section 4E) may represent an additional factor which attaches a follower to his native unit. If the two allegiances cannot both be satisfied, the young hamadryas male seems to give preference to the male clan (see section 7C).

The hamadryas clan is comparable to human clans in the sense that it represents an unilineally determined social unit with spatial localization (Hirschberg 1965). Human clans, however, tend to be exogamous. One function of the endogamy in hamadryas clans might be the optimum preservation of the female lineage system.

The strengthened male-male relationships and the resulting clan structure possibly are not only a compensation for the inhibiting effect of permanent male-female relationships, but may have their own ecological functions: (1) The clan represents an additional level in the multilevel organization of hamadryas and thereby provides additional flexibility for fission and fusion of groups (Kummer 1971). (2) The strengthening of male-male relationships, leading to a high familiarity among males, may enable a degree of cooperation which might be less feasible in a rhesus or Japanese monkey society. The observations of Stolba (in preparation) on Cone Rock troop may provide concrete results to this aspect.

So far, three primate species seem to have strengthened male-male relationships to a degree where they can contribute to the long-term identity of social units: hamadrayas baboons, chimpanzees (Kawanaka and Nishida 1975), and man. A fourth species might be the Guinea baboon (Boese 1975).

The social organization of chimpanzees exemplifies that socially promoted male-male relationships do not necessarily depend on permanent male-female relationships. Chimpanzees form cohorts of adult males which Reynolds (1968) interprets as task forces for exploration and foraging, and which he considers as part of the primate heritage in hominid evolution. Viewed from our framework, two questions about this species remain unanswered: What causes females to leave their native units? What prevents males from leaving their native units, if this suggestion of Kawanaka and Nishida (1975) is confirmed by further studies? Possibly, the differences of a priori compatibilities between males and females are not very marked, so that it might be easier to establish a patrilineally determined society.

In hamadryas the genetic exchange between social units seems to take place through transfers of females. Recent studies on chimpanzees (Kawanaka and Nishida 1975, van Lawick–Goodall 1973) also suggest that females migrate between unit groups. The migration of animals who contribute to the genetic exchange between social units usually appears not to be evenly distributed across a population, rather it is concentrated within local demes. Kawanaka (1973) found that in Japanese monkeys the majority of male migrations occurred within the local concentration of troops which comprises an average of 3 troops living less than 5 kilometers from one another. Kawanaka and Nishida (1975) observed that in chimpanzees two thirds of the recorded female transfers occurred between 2 adjacent unit groups. In human cultures exogamy also tends to occur within the limits of larger social units (Hirschberg 1965). In hamadryas the band appears to be a social unit within which most female transfers between clans take place (tab. 25). These transfers take place on the foraging march; since the clans of a band usually forage within hearing distance females lost by one clan will most probably be taken over by a male of the same band if not of her clan.

D. Mother-Infant Behavior in Socialization

A particular feature of the hamadryas society concerns relationships between mature males and infants. In free-ranging rhesus monkeys males hardly interact with infants (Kaufmann 1966, Lindburg 1971). In the experimental situation only the relationship between an adult male and an infant starts with reactions of withdrawal and hostility and later leads to play and grooming (Redican and Mitchell 1973). Similar behaviors were observed in preadolescent males toward infants (Chamove et al. 1967) and in an adult male toward juveniles (Bernstein and Draper 1964). This sequence of behavior resembles the ontogenetic sequence in peer interactions of rhesus and hamadryas immatures as well as the experimental sequence in adult hamadryas and geladas. In contrast to this sequence interactions between hamadryas males and infants ontogenetically start with mother-infant behavior and proceed to grooming relationships. This corresponds to the sequence within mother-infant dyads. It seems, therefore, that we should distinguish two different bonding sequences, which both may reach the stage of grooming, but which differ in their onset behavior: a play-initiated and a maternal-initiated sequence. So far mother-infant behavior has not yet been integrated into the bonding-stage model because the model was derived from interactions among adult animals. Intuitively it appears plausible to interpret mother-infant behavior as the highest bonding stage: Embracing, its typical behavior, occasionally occurs among unrelated adults, who develop a particularly strong attachment beyond the grooming stage. Accordingly, animals who follow the play-initiated sequence are a priori less compatible than those who follow the maternal-initiated sequence for bond forming.

I have speculated above that high horizontal compatibility is related to the function of long-term identity of social units, low compatibility to the genetic exchange between units through migration. Similarly I should expect that high *vertical* compatibility is also related to the long-term identity of social units, i.e., that if maternal behavior occurs outside mother-infant dyads it should be preferentially performed by animals who contribute to the long-term identity of social units. The available data on primates appear to be in broad agreement with this expectation: In macaques (Bertrand 1969, Chamove et al. 1967, Kaufmann 1966), vervets (Lancaster 1972, Struhsaker 1971) and langurs (Jay 1962, Poirier 1968, Sugiyama 1965a) maternal behavior outside mother-offspring dyads is predominantly performed by females. This behavior has been referred to as "aunt behavior" by Hinde et al. (1964). Maternal behavior is less frequent in males, but if it occurs, it seems to be performed mainly by males who are closely related to the central part of the troop. Itani (1963) reports that in Japanese monkeys male care of infants is predominant in central males; he attributed to these males, besides high sociability and low aggressiveness (= compatibility), a high interest in the central part of the troop. In rhesus males maternal behavior toward infants is particularly rare in peripheral males, and among adults it is more frequent in males who were born in the group than in males who have immigrated (Breuggeman 1973). Regular maternal behavior has been found in baboon males (Morgan and Tuttle 1966, Ransom and Ransom 1971, Rowell et al. 1968), but so far no interpretation has been offered as to what these male-infant interactions contribute to socialization.

Kummer (1967) has interpreted the occurrence of interactions between female infants and subadult males as a transfer of the infant's mother attachment. If the mother-child attachment is considered to be the most compatible bond, its transfer to other animals generally represents an efficient means of socialization. In the hamadryas society it seems to be the basic process of both clan and family forming. the inhibiting effect which the harem bond has on father-offspring interactions (fig. 25) prevents infants from transferring their infantile attachment from their mothers to their fathers, a process which would contribute nothing to socialization. Thus, main secondary caretakers of infants are juvenile and subadult males whose age mates in rhesus or Japanese monkeys already are peripheralized or have left their native troop.

E. Coaction of Dyadic and Triadic Rules

1. Clan and Family Forming

From our observations on clan and family forming there emerges a coaction of dyadic and triadic rules, which represents a rich and well-organized pattern and to me contributes much to the beauty of hamadryas organization. I wish to illustrate how I conceive the overall pattern of this coaction, regardless of how speculative it may be. For the following discussion, reference to figure 25 might be helpful.

Let us start with an intellectual experiment: A subadult hamadryas male has to make his choice among four possible partners, all of which are unfamiliar to him: an adult female, an immature female, an immature male, and a subadult male. If he choses according to his a priori compatibilities with these potential partners, his first choice probably will be the adult female, his second choice the immature female, then the immature male, and, finally, the subadult male. (Note that the preference of the adult to the immature female is not predicted by the present dyadic rules, although it is strongly suggested by qualitative evidence.) This imaginary sequence of partner preference contrasts with the actual situation within the free-ranging band: The other subadult male will already be familiar to him from previous peer contacts; as a young subadult he has established vertical relationships with immature males, and he is now in the onset phase of forming an initial unit with an immature female; he will be a young adult until he has his first adult female. The actual sequence of bond forming thus is the reverse of the above order. The reversal seems to be brought about by an intricate combination of dyadic compatibilities and triadic effects: Males, who would form dyads of low compatibility as subadults, become familiar with one another in immature peer interactions when they are still compatible. Play among immatures occurred in agreement with a priori compatibilities. The tendency of immature males to interact increasingly with immature females, which should be expected from the dyadic rules, is suppressed by the establishment of MF-initial bonds, which make immature females less accessible to immature males (tab. 24). Harem bonds, father-daughter bonds, and MF-initial bonds make it difficult for the growing hamadryas male to interact with females of the same or older age classes. Increasing age, however, permits him compatible relationships with infants; these relationships are shaped as maternal-initiated, not as play-initiated sequences of bonding. Interactions between bachelors and infants are further facilitated by the effect of the harem bond, which suppresses interactions between harem leaders and their infants; this triadic effect might prevent bachelors from competing overtly with harem leaders over access to infants. From the dyadic rules we should expect bachelors to prefer mothering female rather than male infants; actually they preferentially take care of male infants (fig. 29). Two factors may account for this: Firstly, male infants seem more easily accessible to bachelors, because they more often leave their parental unit to form play groups; secondly, the father-daughter bond appears to have some inhibiting effect on the bachelor's behavior toward immature females.

So far, the social ontogeny of a growing hamadryas male seems not to differ markedly from the development of a rhesus male. During the juvenile period rhesus males also show increasingly maternal behavior toward infants, reaching a maximum of frequency in 3-year-old males (Breuggeman 1973). Rhesus males more frequently mother male infants, particularly during the mating season (Breuggeman 1973), although under experimental conditions rhesus males show more friendly behavior toward female infants (Brandt and Mitchell 1973, Stevens and Mitchell 1972). The main difference between rhesus and hamadryas males appears to be that interactions between males and infants do not lead to lasting relationships in rhesus monkeys. Although Breuggeman (1973) recognizes a tendency for male-male bonds to form via maternal care, the decrease of maternal behavior in 4- and 5-year-old males coincides with the mean age of emigrating males (Drickamer and Vessey 1973). In subadult hamadryas males the decrease of maternal behavior (figs. 10 and 11) is not the end of the relationship but occurs when interactions between subadult males and infants are transformed to grooming relationships.

In Japanese monkeys, Itani (1963) found 28 males among 68 one-year-old infants who were mothered by males, but only 5 males among 25 two-year-old infants. He explained this change of proportion with the increasing tendency of immature males to become peripheralized, a process which makes immature males less accessible to the caretaking central males. Although Imanishi (1965) suggested that interactions between central males and male infants may increase the infant's probability to remain in the center and there to become a leader, the process seems to be exception rather than the rule.

Thus, the formation of vertical male-male relationships via maternal behavior occurs only as a tendency in the matrilineal societies of rhesus and Japanese monkeys, whereas hamadryas base their patrilineal clan structure on this preadaptation. In rhesus and Japanese monkeys this process is interrupted by the peripheralization and isolation of males; males are peripheralized and leave their native troop because they cannot establish compatible relationships with central males.

In hamadryas, the permanent harem bond, on one hand, requires the strengthening of male-male relationships since it suppresses interactions between possessors of females and other males; on the other hand, it contributes to the promotion of male-male relationships by suppressing the more compatible interactions between bachelors and adult females. An elegant solution indeed.

2. The Ontogenic Transition from Male-Male to Male-Female Relationships

Male-male relationships are promoted by suppressing male-female interactions. What factors ensure that males, after having predominantly interacted with partners of their own sex, establish relationships with females? Is it not possible that males become completely attached to male partners? Actually no male has been found during our study who did not acquire females. The male who appeared most attached to male partners was Steward in band III; but even Steward acquired an initial female and, later, established a one-male unit.

The main factor which promotes the transition from the predominance of male-male interactions to family forming seems to be the great difference of compatibility between male-male and male-female interactions. Even a highly familiar relationship between two males appears less compatible than a completely unfamiliar male-female pair. This is supported by the experiments of Kummer et al. (1974), who brought together two males from the same troop and a female from an alien troop. There is an important difference between this experimental situation and the conditions in a free-ranging band: While the rivals in the experiments had no other partner for social interactions, bachelors in a free-ranging band can interact with one another. These male-male relationships, however, do not abolish their interest in females: The frequencies of notifying at the sleeping cliff suggested that rivalry for females remained a typical characteristic between harem leaders and bachelors (section 7A). In addition, table 36 shows that in the 26 close-study males the frequency of conflict behaviors was negatively correlated to grooming and mother-infant interactions with female partners and positively to such interactions with male partners. Kummer et al. (1974) interpreted self-grooming, scratching, and wiping muzzle as produced by a conflict between the rival's attraction to the pair on one hand, and escape and social inhibition on the other hand; moreover, they suggested that self-grooming may be redirected grooming, a suggestion which also is consistent with observations on macaques (Goosen 1974, Rosenblum et al. 1966). The figures of table 36 are compatible with the view that male-male relationships are not an adequate substitute for male-female bonds: frequent grooming among males does not reduce the frequency of self-grooming. This interpretation suggests that the fre-

Table 36. Correlation of Conflict Behaviors to Grooming and Mother-Infant Interactions with Male and Female Partners in the Close-Study Males

Conflict behaviors	Correlation with interactions between	
	Males and females	Males and males
Self-grooming	−0.56**	+0.46*
Scratching	−0.33*	+0.14
Yawning	−0.49**	+0.50**
Wiping muzzle	−0.52**	+0.27
Shaking body	−0.43*	+0.29

The sample comprises 26 males in age classes N to young adult (table 2) and covers a total observation time of 137 hours. The frequencies of conflict behaviors are correlated with frequencies of male-male and male-female interactions (Spearman rank correlation). Interactions include only grooming and mother-infant behavior, which are considered to be the highest bonding stages. Male-male and male-female interactions are negatively correlated to one another with $R = -0.72**$. * indicates a $p < 0.05$, ** a $p < 0.001$.

quency of grooming alone is not a sufficient measure of compatibility, and that a measure which combines grooming and self-grooming might be more appropriate.

Using a chemical terminology, one may summarize the coaction of dyadic rules and triadic effects as follows: The sex- and status-dependent dyadic compatibilities determine which sex-age class representatives are most likely to form stable combinations. Each individual probably tends to form its social relationships according to the primary pattern of a priori compatibilities, i.e., it tends to maximize its "sum of realized compatibilities," as Kummer (1975) has suggested for geladas. Triadic effects work as catalysts which moderate bond forming and, in particular, promote the occurrence of bonds that would otherwise be unlikely.

Two essential features are not described by the chemical analogy: (1) Unlike chemical affinity, which depends on the stable structures of atoms, a priori compatibilities with unfamiliar partners depend on factors which change throughout the life time of individuals. This makes the ontogenetic point of time at which a relationship is started an important factor for socialization. In general, the greater the compatibility which a society requires of one of its relationships, the earlier it should allow it to start. (2) Social bond forming, unlike chemical synthesis, is an irreversible process: its result, familiarity, is not easily lost. Familiarity proves here to be essential for an understanding of socialization. Theoretically, the ontogenetic point in time would be of little importance if bond forming did not produce a time-resistant result: familiarity. It seems plausible to attribute the following characteristics to familiarity: (a) Familiarity resists to the duration of separation of individuals and permits a higher compatibility between reunited familiar animals than between newly convened unfamiliar ones. Erwin et al. (1975) paired familiar rhesus monkeys who had been separated for 2 years and compared their behavior to unfamiliar dyads: the familiar dyads generally showed less agonistic behavior and higher proximity scores. (b) Familiarity resists to the effect of increasing status, i.e., of age on an ontogenetic scale. Thus two familiar animals of a given sex-age combination should have a more compatible relationship than two unfamiliar animals of the same sex-age combination. We have used this presumed characteristic as an axiom when testing the dyadic rules (section 4D). Later, we found it confirmed by mature males who did not become involved in the gradual development of initial units: They did not groom male partners less frequently than did juvenile males (fig. 11 and tab. 22), although their increasing status should have made them less compatible in the meantime. Kummer's (1975) experiments on geladas also suggest that high familiarity generally results in a higher compatibility of an established relationship. (c) Familiarity offers some resistance to triadic effects. A familiar relationship should maintain a higher compatibility in the face of inhibiting effects than does an unfamiliar relationship. This presumed characteristic still lacks experimental testing. The experiments of Kummer et al. (1974) provide no conclusive results because males of the same troop might not have been familiar enough; they might have been from different bands or from different clans within a band. We have used the presumed resistance of familiar relationships to the inhibition by triadic effects to explain why leader-follower associations are maintained in spite of the inhibiting effect of harem bonds (section 7E). It agrees with the observation that grooming within MM-initial bonds persists in the face of female possession: In band I Minimus still repeatedly groomed Cadet after Cadet had established his initial unit containing three juvenile females; even after the breakup of Padorn's one-male unit, which provided Cadet with an adult female, grooming was not given up completely by Cadet and Minimus. In 1973 and 1974, however, when Cadet had enlarged his one-male unit and when Minimus was on the way to establish his own initial unit, grooming was not any more observed between the two males. A similar development was observed between Bishop and Pepsi. These two males stopped grooming only after Bishop had enlarged his one-male unit and Pepsi's initial bond was reaching the stage of full antonomy. In band III, Steward and Garu (see section 7D) still groomed one another after Steward had acquired an initial female. Kummer (1968a, p. 67) observed regular grooming between two initial leaders and interpreted it as a persisting relationship carried over from the bachelor period. The observations of two males (section 7G) suggested that if he has to leave his native clan, a male's integration into the band organization is poor and the establishment of his initial unit is delayed; apparently, such males lack the high familiarity with the males of the host clan, which, in the native clan, is usually attained during early ontogeny. *(d)* Familiarity offers some resistance to the formation of new bonds. This presumed effect of familiarity was already suggested by the experiments of Kummer (1975) on geladas; it also agrees with our observation that transferred females, who have established familiar relationships in their original one-male unit, were reluctant in forming relationships with females of the new one-male unit (section 6B). Theoretically, the resistance to the forming of new bonds is important because it prevents an individual from engaging in more and more compatible relationships, and it contributes to the difference of behavior toward familiar group members and toward strangers. Since familiarity develops gradually during ontogeny we should expect that play, which expresses a low degree of familiarity, occurs within a large range of individuals, whereas grooming should occur within smaller parties of familiar animals whose relationships constitute structures of the social organization. This actually appears to be the case in Japanese monkeys (Yamada 1963) and in rhesus (Loy and Loy 1974, Sade 1963): in these species play seems less restricted to matrilineally related partners than grooming. In hamadryas it agrees with the result that the preference of males for interacting with males of their own clan is stronger for grooming than for nongrooming interactions, among them of notifying. The resistance to new relationships, however, is overridden if the prospective new relationship has a markedly greater a priori compatibility than the familiar relationship.

Thus, a primatological theory on socialization should include the combination of two sets of factors: Firstly, the combination of a priori compatibilities which determine the primary course for social dynamics, and of triadic (possibly also polyadic) effects which modify the primary pattern of a priori compatibilities. Secondly, the combination of age-dependent compatibilities, which make the time of starting a relationship an important variable for socialization, and of familiarity, which represents a time-resistant result of bond-forming. These two combinations of factors are related to one another: The coaction of dyadic and triadic rules determines the ontogenetic sequence of starting relationships and thus contributes to the pattern of familiar relationships which constitute important structures of social organization.

F. A Glimpse at Man

Ethnologists and sociologists might greatly contribute to a differentiation and refinement of the framework presented in this chapter.

Man belongs to the few primate species which make use of male-male relationships for the long-term identity of social units. Almost all human cultures have permanent male-female relationships in the sense that particular individuals are assigned to each other through marriage rites; and almost all human cultures have all-male societies, which appear more pronounced and institutionalized than all-female groupings. These all-male societies are not simply temporary reservoirs for peripheralized males, but bring together males whose relationships are to become part of important social structures. In contrast to nonhuman primates, human all-male groups contain not only bachelors but married men as well, probably because exclusive relationships among humans are much less dependent on permanent spatial proximity than the bonds of hamadryas baboons. Human societies can thus combine the hamadryas way of strengthening male-male relationships with that of rhesus or Japanese macaques, i.e., they can establish bachelor bonds which exert their after-effect during married life, and they can maintain active and intense relationships among married men such as we find among rhesus or Japanese macaque males during the nonmating period. We should particularly expect that men who are to occupy high status positions tend to attain mutual familiarity within all-male societies since a high degree of cooperation is especially important among high-ranking males while, at the same time, unfamiliar males of high ranks are particularly incompatible. Furthermore, it would be of comparative interest to know whether the particular organization and composition of a human all-male group is related to particular social and economic factors.

Kin relations play a significant role among humans as well. However, the terms *patrilineal* and *matrilineal* seem to have different meanings in primatology and ethnology. When primatologists use the terms, they refer to genetic relatedness; to the ethnologist, kinship as seen by the people described is a social convention which is related to allocation of status and from which rights and obligations with regard to inheritance, location after marriage, and family formation are derived. Inherent in both views is the concept of familiarity (note that *familiarity* etymologically does not imply the notion of kinship, but is derived from the totality of famuli, of domestic servants living in the same house community). Kinship, whether genetic or not, creates (and results from) familiar relationships that become factors of social structure. Because of this common aspect, one may ask whether it is possible to recognize compatibility-dependent features even in human kinship systems. Thus, it seems difficult to combine a female lineage system with permanent male-female relationships and pronounced male-male cooperation within one social organization, because permanent male-female bonds negatively affect male-male interactions (although in humans this effect may be compensated by all-male societies which include married males, see above), and because a pronounced cooperation among males seems best warranted by a male kinship system, which tends to disintegrate the female lineage system. Therefore, matrilineally-oriented human cultures deserve particular attention. The following predictions appear to be compatible with our concept.

Matrilineally oriented cultures are less frequent than patrilineally or bilineally oriented cultures and occur preferentially under economic conditions which do not require a pronounced cooperation among males. Male-female relationships might show features which can be interpreted as a weakening of the bond; this weakening can be institutionalized in that husband and wife keep separate residences after marriage, a feature which actually seems to occur preferentially in matrilineally oriented cultures (Hirschberg 1965); it can appear as a reduced authority of the father toward his children in favor of an increased authority of the mother's brother, a feature which seems characteristic for matrilineally oriented cultures as well (Hirschberg 1965); the weakening of male-female bonds may manifest itself in a high rate of divorce.

Status allocation in the sense of ethnology could become valuable to primatological theory. It is possible that status affects the process of socialization. Laboratory primatologists usually determine dominance status by observing the interactions in isolated dyads; they thereby neglect possible effects of the animals' life histories. Instead, they search for correlations of status with such individual characters as body weight (Norikoshi and Koyama 1975) or hormonal state (Joslyn 1973). This restricted concept of status also lay at the basis of the bonding-stage model: Kummer (1975) described compatibility in dependence of status and assumed status to be a single variable such as a correlate of size or of masculinity. Kawai (1965) was the first primatologist who not only correlated status with ahistoric individual qualities but analyzed status as a result of socialization processes. He differentiated between a basic rank, which manifests between two animals in the absence of a third party, and a dependent rank, which is affected by social factors such as kinship or spatial location in the troop. He argues that social factors take effect early in ontogeny in that the mother and other relatives influence the course of the infant's dyadic interactions and thus establish even the latter's basic rank for purely dyadic situations. The effect is so marked that it leads to a powerful order in the genealogical and cross-genealogical status relations of Japanese monkeys (Kawai 1965, Kawamura 1965, Koyama 1976) and of rhesus (Loy and Loy 1974, Missakian 1972). These findings of longitudinal studies of primate species accentuate the similarity between nonhuman and human kinship as a status-allocating or status-affecting system.

9 Summary

Some aspects of socialization in hamadryas baboons have been investigated during an eighteen-month field study in Ethiopia. Behavioral samples were taken from marked and unmarked individuals and censuses from three bands of a troop. The analysis provided the following results:

(1) The *band* is a stable unit within the troop. It is an autonomous unit with respect to foraging and to the processes of socialization. Male transfers between bands appear to be less frequent than in other cercopithecid species; possibly, female transfers across bands contribute to the genetic exchange between bands.

(2) Close study of one band revealed a new type of subunit: the clan. A *clan* comprises a few family units which differ from each other in their developmental stage and which are kept together by a quasilinear age hierarchy of males. Male infants become integrated into the clan by bachelors who kidnap and mother them. These interactions presumably persist as leader-follower associations.

(3) *Families* are formed within the clan. First, the initial unit develops in that bachelors kidnap and mother female infants and later separate them from the parental one-male unit. This separation is subtle and gradual or sudden, apparently depending on the particular band. The mature one-male unit is developed in that the initial leader defeats the oldest harem leader of his clan; in the resulting breakup the adult females of the dissolved unit are taken over by the initial leaders, not by other dominant harem leaders.

(4) *Sex-specific differences* in behavior occur already among immatures. Immature males tend to show more aggressive, females more friendly behavior.

An attempt was made at integrating the results of hamadryas socialization into the bonding-stage model of Kummer (1975). This model describes bond forming as a *sequence* of behavior, of *dyadic compatibility* in dependence of status, and of the influence of *triadic effects* on bond forming. It permits the following interpretation of the results: The permanent harem bonds suppress interactions between female possessors and other males, and they tend to disintegrate the female lineage system which warrants the long-term identity of social units in macaque societies. Harem bonds, initial bonds, and father-daughter bonds suppress the highly compatible interactions between bachelors and females and thus promote the occurrence of the less compatible male-male interactions. These socially promoted male-male relationships are established before they are affected by the possession of females; the resulting familiarity among males presumably prevents their isolation; it furthers the resistance of their relationships against the inhibiting effect of male-female bonds in their later life, and it establishes the male clan system, which contributes to the long-term identity of social units.

References

Altmann, J. 1974. "Observational Study of Behavior: Sampling Methods." *Behaviour* 49: 227–67.

Altmann, S. A., and Altmann, J. 1970. "Baboon Ecology: African Field Research." *Bibl. Primat.*, vol. 12. Basel: Karger.

Baldwin, J. D. 1969. "The Ontogeny of Social Behaviour of Squirrel Monkeys *(Saimiri sciureus)* in a Seminatural Environment." *Folia Primat.* 11: 35–79.

Baldwin, J. D., and Baldwin, J. I. 1973. "The Role of Play in Social Organisation." *Primates* 14: 369–81.

Bernstein, I. S., and Draper, W. A. 1964. "The Behaviour of Juvenile Rhesus Monkeys in Groups." *Animal Behaviour* 12: 84–91.

Bertrand, M. 1969. "The Behavioral Repertoire of the Stumptail Macaque." *Bibl. Primat.*, vol. 11 Basel: Karger.

Boelkins, R. C., and Wilson, A. P. 1972. "Intergroup Social Dynamics of the Cayo Santiago Rhesus *(Macaca mulatta)* with Special Reference to Changes in Group Membership by Males. *Primates* 13: 125–40.

Boese, G. K. 1975. "Social Behavior and Ecological Considerations of West African Baboons *(Papio papio)*." In Tuttle, *Socioecology and Psychology of Primates*, pp. 205–30. The Hague: Mouton.

Borges, J. L. 1964. *Labyrinths*. New York: New Directions.

Brandt, E. M., and Mitchell, G. 1973. "Pairing Preadolescents with Infants *(Macaca mulatta)*." *Dev. Psychol.* 8, 222–28.

Breuggeman, J. A. 1973. "Parental Care in a Group of Free-Ranging Rhesus Monkeys *(Macaca mulatta)*." *Folia Primat.* 20: 178–210.

Burton, F. D. 1972 "The Integration of Biology and Behavior in the Socialization of *Macaca sylvana* in Gibraltar." In Poirier, *Primate Socialization*, pp. 29–62. New York: Random House.

Chamove, A. S.; Harlow, H. F.; and Mitchell, G. 1967. "Sex Differences in the Infant-Directed Behavior of Preadolescent Rhesus Monkeys." *Child Dev.* 38: 329–35.

Chance, M. R. A., and Jolly, C. J. 1970. *Social Groups of Monkeys, Apes and Men*. London: Jonathan Cape.

Drickamer, L. C. 1974. "A Ten-Year Summary of Reproduction Data for Free-Ranging *Macaca mulatta*." *Folia Primat.* 21: 61–80.

Drickamer, L. C., and Vessey, S. H. 1973. "Group Changing in Free-Ranging Male Rhesus Monkeys." *Primates* 14: 359–68.

Eisenberg, J. F.; Muckenhirn, N. A.; and Rudran, R. 1972. "The Relation between Ecology and Social Structure in Primates." *Science* 176: 863–74.

Erwin, J.; Maple, T.; and Welles, J. F. 1975. "Responses of Rhesus Monkeys to Reunion." *Contemporary Primatology. 5th Int. Congr. Primat. Nagoya 1974*, pp. 254–62. Basel: Karger.

Furuya, Y. 1968. "On the fission of Troops of Japanese Monkeys. I. Five fissions and Social Changes between 1955 and 1966 in the Gagyusan Troop." *Primates* 9: 323–50.

Gartlan, J. S., and Brain, C. K. 1968. "Ecology and Social Variability in *Cercopithecus aethiops* and *C. mitis.*" In Jay, *Primates: studies in Adaptation and Variability*, pp. 253–92. New York: Holt, Rinehart and Winston.

Goosen, C. 1974. "Immediate Effects of Allogrooming in Adult Stump-tailed Macaques, *Macaca arctoides.*" *Behavior* 48: 75–88.

Hall, K. R. L., and Devore, I. 1965. "Baboon Social Behavior." In Devore, *Primate Behavior*, pp. 53–110. New York: Holt, Rinehart and Winston.

Hanby, J. P., and Brown, C. E. 1974. "The Development of Sociosexual Behaviours in Japanese Macaques, *Macaca fuscata.*" *Behaviour* 49:152–96.

Hansen, E. W. 1966. "The Development of Maternal and Infant Behavior in the Rhesus Monkey." *Behaviour* 27: 107–49.

Harlow, H. F. 1966. "The Primate Socialization Motives." *Transactions and Studies of the College of Physicians of Philadelphia* 33: 224–37.

Harlow, H. F., and Harlow, M. K. 1965. "The Affectional Systems." In Schrier, Harlow, and Stollnitz, *Behavior of Nonhuman Primates*, 2:287–334. New York: Academic Press.

Hausfater, G. 1972. "Intergroup Behavior of Free-Ranging Rhesus Monkeys (*Macaca mulatta*)." *Folia primat.* 18:78–107.

Hinde, R. A.; Rowell, T. E., and Spencer-Booth, Y. 1964. "Behaviour of Socially Living Rhesus Monkeys in Their First Six Months." *Proc. Zool. Soc. London* 143:609–49.

Hinde, R. A., and Spencer-Booth, Y. 1967. "The Behaviour of Socially Living Rhesus Monkeys in Their First Two and a Half Years" *Animal Behaviour* 15:169–96.

Hinderling, R. 1975. "Die Entwicklung von Zweierbeziehungen beim Mantel-pavian." Unpubl. diploma thesis, University Zurich.

Hirschberg, W. 1965. *Wörterbuch der Völkerkunde*. Stuttgart: Alfred Körner Verlag.

Imanishi, K. 1965. "Identification: A Process of Socialization in the Subhuman Society of *Macaca fuscata.*" In Altmann, *Japanese Monkeys*, pp. 30 51. Edmonton: University of Alberta Press.

Itani, J. 1963. "Paternal Care in the Wild Japanese Monkey, *Macaca fuscata.*" In Southwick, *Primate Social Behavior*, pp. 91–97. Princeton, N.J.: Van Nostrand.

Itani, J.; Tokuda, K.; Furuya, Y.; Kano, K.; and Shin, Y. 1963. "The Social

Construction of Natural Troops of Japanese Monkeys in Takasakiyama." *Primates* 4 (3): 1–42.

Jay, P. C. 1962. "Aspects of Maternal Behavior among Langurs." *Ann. N.Y. Acad. Sci.* 102: 468–76.

———. 1965. "The Common Langur of North India." In Devore, *Primate Behavior*, pp. 197–249. New York: Holt, Rinehart and Winston.

Jolly, A. 1972. *"The Evolution of Primate Behavior."* New York: Macmillan.

Joslyn, J. W. 1973. "Androgen-Induced Social Dominance in Infant Female Rhesus Monkeys." *J. Child Psychol. Psychiat.* 14: 137–45.

Kaufmann, J. H. 1965. "A Three-Year Study of Mating Behavior in a Free-Ranging Band of Rhesus Monkeys." *Ecology* 40: 500–512.

———. 1966. "Behavior of Infant Rhesus Monkeys and their Mothers in a Free-Ranging Band." *Zoologica* 51:17–28.

Kaufman, I. C., and Rosenblum, L. A. 1967. "Depression in Infant Monkeys Separated from Their Mothers." *Science* 155:1030–31.

Kawai, M. 1965. "On the System of Social Ranks in a Natural Troop of Japanese Monkeys." I.: "Basic Rank and Dependent Rank." In Altmann, *Japanese Monkeys*, pp. 66–86. Edmonton: University of Alberta Press.

Kawamura, S. 1965. "Matriarchal Social Ranks in the Minoo-B Troop: A Study of the Rank System of Japanese Monkeys." In Altmann, *Japanese Monkeys*, pp. 105–12. Edmonton: University of Alberta Press.

Kawanaka, K. 1973. "Intertroop Relationships among Japanese Monkeys." *Primates* 14: 113–59.

Kawanaka, K., and Nishida, T. 1975. "Recent Advances in the Study of Inter-unit-group Relationships and Social Structure of Wild Chimpanzees of the Mahali Mountains." *Proc. Symp. 5th Congr. Int. Primat. Soc.*, pp. 173–86. Tokyo: Japan Science Press.

Koford, C. B. 1966. "Population Changes in Rhesus Monkeys: Cayo Santiago, 1960–1964." *Tulane Studies in Zoology* 13:1–7.

Koyama, N. 1967. "On Dominance Rank and Kinship of a Wild Japanese Monkey Troop in Arashiyama." *Primates* 8:189–216.

———. 1970. "Changes in Dominance Rank and Division of a Wild Japanese Monkey Troop in Arashiyama." *Primates* 11:335–90.

Krech, D.; Crutchfield, R. S.; and Ballachey, E. L. 1962. *"Individual in Society."* New York: McGraw-Hill.

Kummer, H. 1967. "Tripartite Relations in Hamadryas Baboons." In Altmann, *Social Communication among Primates*, pp. 63–71. Chicago: University Chicago Press.

———. 1968a. "Social Organisation of Hamadryas Baboons: A Field Study." *Bibl. primat.*, vol. 6. Basel: Karger.

———. 1968b. "Two Variations in the Social Organization of Baboons." In Jay, *Primates: Studies in Adaptation and Variability*, pp. 293–312. New York: Holt, Rinehart and Wisnton.

——. 1971a. *Primate Societies: Group Techniques of Ecological Adaptation.* Chicago: Aldine, Atherton.

——. 1971b. "Immediate Causes of Primate Social Structures." *Proc. 3rd Int. Congr. Primat.* Zurich 1970. 3:1–11. Basel: Karger.

——. 1975. "Rules of Dyad and Group Formation among Captive Gelada Baboons *(Theropithecus gelada). Proc. Symp. 5th Congr. Int. Primat. Soc.*, pp. 129–59. Tokyo: Japan Science Press.

Kummer, H.; Goetz, W.; and Angst, W. 1970. "Cross-Species modification of Social Behavior in Baboons." In Napier and Napier, *Old World Monkeys: Evolution, Systematics and Behavior*, pp. 351–63. London: Academic Press.

Kummer, H.; Goetz, W.; and Angst, W. 1974. "Triadic Differentiation: An Inhibitory Process Protecting Pair Bonds in Baboons." *Behaviour* 49:62–87.

Lancaster, J. B. 1972. Play-Mothering: The Relations between Juvenile Females and Young Infants among Free-Ranging Vervet Monkeys." In Poirier, *Primate Socialization*, pp. 83–104. New York: Random House.

Lawick, Goodall, J. van. 1967. "Mother-Offspring Relationship in Wild Chimpanzees." In Morris, *Primate Ethology*, pp. 287–346. London: Weidenfield and Nicolson.

——. 1973. "The Behaviour of Chimpanzees in Their Natural Habitat." *Am. J. Psychiat.* 130:1–12.

Lindburg, D. G. 1969. "Rhesus Monkeys: Mating Season Mobility of Adult Males." *Science* 166:1176–78.

——. 1971. "The Rhesus Monkey in North India: An Ecological and Behavioral Study." In Rosenblum, *Primate Behavior*, 2:1–106. New York: Academic Press.

Loy, J., and Loy, K. 1974. "Behavior of an All-Juvenile Group of Rhesus Monkeys. *Am. J. Phys. Anthrop.* 40: 83–95.

Mason, W. A. 1963. "The Effects of Environmental Restriction on the Social Development of Rhesus Monkeys. In Southwick, *Primate Social Behavior*, pp. 161–73. Princeton, N.J.: Van Nostrand.

——. 1965a. "Determinants of Social Behavior in Young Chimpanzees." In Schrier, Harlow, and Stollnitz, *Behavior of Nonhuman Primates*, pp. 335–64. New York: Academic Press.

——. 1965b. "The Social Development of Monkeys and Apes." In DeVore, *Primate Behavior*, pp. 514–43. New York: Holt, Rinehart and Winston.

Miller, M. H.; Kling, A.; and Dicks, D. 1973. "Familial Interactions of Male Rhesus Monkeys in a Semi-Free-Ranging Troop. *Am. J. Phys. Anthrop.* 38:605–12.

Missakian, E. A. 1972. "Genealogical and Cross-Genealogical Dominance Relations in a Group of Free-Ranging Rhesus Monkeys *(Macaca mulatta)* on Cayo Santiago." *Primates* 13:169–80.

Mitchell, G. D. 1968. Persistent Behavior Pathology in Rhesus Monkeys following Early Social Deprivation." *Folia Primat.* 8:132–47.

Mitchell, G.; Ruppenthal, G. C.; Raymond, E. J.; and Harlow, H. F. 1966. "Long-Term Effects of Multiparous and Primiparous Monkey Mother Rearing. *Child Develop.* 37:781–91.

Morgan, M. T., and Tuttle, R. H. 1966. "Intimate Infant–Adult Male Interactions in Rhodesian Baboons *(Papio cynocephalus)*." *Am. J. Phys. Anthrop.* 25:203.

Nagel, U. 1973. "A Comparison of Anubis Baboons, Hamadryas Baboons and Their Hybrids at a Species Border in Ethiopia." *Folia Primat.* 19:104–65.

Neville, M. K. 1968. "A Free-Ranging Rhesus Troop Lacking Adult Males. *J. Mammal.* 49:771–73.

Nishida, T. 1966. "A Sociological Study of Solitary Male Monkeys." *Primates* 7:141–204.

Nishida, T., and Kawanaka, K. 1972. "Inter-Unit-group Relationships among Wild Chimpanzees of the Mahali Mountains." In Umesao, *Kyoto University African Studies,* 7:131–69. Kyoto: Kyoto University Press.

Norikoshi, K., and Koyama, N. 1975. "Group Shifting and Social Organization among Japanese Monkeys." *Proc. Symp. 5th Congr. Int. Primat. Soc.,* pp. 43–61. Tokyo: Japan Science Press.

Pfanzagl, J. 1968. *Allgemeine Methodenlehre der Statistik II. Sammlung Göschen Band 747.* Berlin: Walter de Gruyter.

Poirier, F. E. 1968. "The Nilgiri Langur *(Presbytis johnii)* Mother-Infant Dyad." *Primates* 9:45–68.

———. 1969. "The Nilgiri Langur *(Presbytis johnii):* "Its Composition, Structure, Function and Change." *Folia Primat.* 10: 20–47.

———. 1972. *Primate Socialization.* New York: Random House.

Ransom, T. W., and Ransom, B. S. 1971. "Adult Male–Infant Relations among Baboons *(Papio Anubis)*." *Folia Primat.* 16: 179–95.

Ransom, T. W., and Rowell, T. E. 1972. "Early Social Development of feral Baboons." In Poirier, *Primate Socialization,* pp. 105–44. New York: Random House.

Redican, W. K., and Mitchell, G. 1973. "A Longitudinal Study of Paternal Behaviour in Adult Rhesus Monkeys." *Dev. Psychol.* 8: 135–36.

Reynolds, V. 1968. "Kinship and the Family in Monkeys, Apes and Man." *Man* 3:209–23.

Rosenblum, L. A. 1968. "Mother-Infant Relations and Early Behavioral Development in the Squirrel Monkey." In Rosenblum and Cooper, *The Squirrel Monkeys,* pp. 207–34. New York: Academic Press.

———. 1971. "Infant Attachment in Monkeys." In Schaffer, *Origins of Human Social Relations,* pp. 85–113. New York: Academic Press.

Rosenblum, L. A.; Kaufman, I. C.; and Stynes, A. J. 1966. "Some Characteristics of Adult Social and Autogrooming Patterns in Two Species of Macaque." *Folia Primat.* 4:438–51.

Rosenblum, L. A., and Kaufman, I. C. 1967. "Laboratory Observations of

Early Mother-Infant Relations in Pigtail and Bonnet Monkeys." In Altmann, *Social Communication among Primates*, pp. 33–41. Chicago: University of Chicago Press.

Rowell, T. E. 1966. "Forest Living Baboons in Uganda." *J. Zool. Lond.* 149:344–64.

———. 1969. "Long-Term Changes in a Population of Uganda Baboons." *Folia Primat.* 11: 241–54.

Rowell, T. E.; Din, N. A.; and Omar, A. 1968. "The Social Development of Baboons in Their First Three Months." *J. Zool. Lond.* 155:461–83.

Rudran, R. 1973. "Adult Male Replacement in One-Male Troops of Purple-Faced Langurs *(Presbytis senex senex)* and Its Effect on Population Structure." *Folia Primat.* 19: 166–92.

Saayman, G. S. 1968. "Oestrogen, Behaviour and Permeability of a Troop of Chacma Baboons." *Nature* 220:1339–40.

Sade, D. S. 1963. "Ontogeny of Social Relations in a Group of Free-Ranging Rhesus Monkeys." Ph. D. dissertation, University of California, Berkeley.

———. 1965. "Some Aspects of Parent-Offspring and Sibling Relations in a Group of Rhesus Monkeys with a Discussion of Grooming." *Am. J. Phys. Anthrop.* 23:1–18.

———. 1968. "Inhibition of Son-Mother Mating among Free-Ranging Rhesus Monkeys." *Sci. Psychoanal.* 12: 18–37.

———. 1972. "A Longitudinal Study of Social Behavior of Rhesus Monkeys." In Tuttle, *The Functional and Evolutionary Biology of Primates*, pp. 378–98. Chicago: Aldine-Atherton.

Seay, B.; Alexander, B. K.; and Harlow, H. F. 1964. "Maternal Behavior of Socially Deprived Rhesus Monkeys. *J. Abnormal a. Soc. Psychol.* 69:345–54.

Simonds, P. E. 1965. "The Bonnet Macaque in South India." In DeVore, *Primate Behavior*, pp. 175–96. (Holt, Rinehart and Winston, New York 1965).

———. 1973. "Outcast Males and Social Structure among Bonnet Macaques *(Macaca radiata)*." *J. Am. Phys. Anthropol.* 38:599–604.

Spencer-Booth, Y. 1969. "The Effects of Rearing Rhesus Monkey Infants in Isolation with Their Mothers on Their Subsequent Behaviour in a Group Situation."

Spencer-Booth, Y., and Hinde, R. A. "Effects of six-day Separation from Mother on 18- to 32-Week-Old Rhesus Monkeys." *Animal Behaviour* 19:174–91.

Spiegel, A. 1954. "Beobachtung und Untersuchung an Javamakaken." *Zool. Garten* 20:227–70.

Stevens, C. W., and Mitchell, G. 1972. "Birth Order Effects, Sex Differences and Sex Preferences in the Peer-Directed Behavior of Rhesus Infants." *Int. J. Psychobiol.* 2:117–28.

Stolz, L. P. 1972. "The Size, Composition and Fissioning in Baboon Troops *(Papio ursinus)*." *Zoologica Africana* 7:367–78.

Struhsaker, T. T. 1967. "Social Structure among Vervet Monkeys *(Cercopithecus aethiops)*." *Behaviour* 29:83–121.

———. 1971. T. T.: "Social Behaviour of Mother and Infant Vervet Monkeys *(Cercopithecus aethiops)*." *Animal Behaviour* 19:233–50.

Sugiyama, Y. 1965a. "Behavioral Development and Social Structure in Two Troops of Hanuman Langurs *(Presbytis entellus)*." *Primates* 6:213–47.

———. 1965b. "On the Social Change of Hanuman Langurs *(Presbytis entellus)* in Their Natural Condition." *Primates* 6:381–418.

———. 1967. "Social Organization of Hanuman Langurs. In Altmann, *Social Communication among Primates*, pp. 221–36. Chicago: University of Chicago Press.

Tokuda, K. 1962. "A Study on the Sexual Behavior in the Japanese Monkey Troop." *Primates* 3:1–40.

Vessey, S. H. 1971. "Free-Ranging Rhesus Monkeys: Behavioural Effects of Removal, Separation and Reintroduction of Group Members." *Behaviour* 40:216–27.

Wickler, W. 1967. "Sociosexual Signals and Their Intraspecific imitation among primates. In Morris, *Primate Ethology*," pp. 69–147. London: Weidenfield and Nicolson.

Yamada, M. 1963. "A Study of Blood-Relationship in the Natural Society of the Japanese Macaque." *Primates* 4:43–65.

———. 1971. "Five Natural Troops of Japanese Monkeys on Shodoshima Island. II. A Comparison of Social Structure." *Primates* 12:125–50.

Yoshiba, K. 1968. "Local and Intertroop Variability in Ecology and Social Behavior of Common Indian Langurs." In Jay, *Primates: Studies in Adaptation and Variability*, pp. 217–42. New York: Holt, Rinehart and Winston.

Subject Index